# LAST NIGHT
## in the OR

A Transplant Surgeon's Odyssey

## BUD SHAW, MD

A PLUME BOOK

PLUME
An imprint of Penguin Random House LLC
375 Hudson Street
New York, New York 10014
penguin.com

"Good Opera" adapted from Bud Shaw, "My Night with Ellen Hutchinson," *Creative Nonfiction* 42 (2011): 23–27.

LIBRARY OF CONGRESS CATALOGING-IN-PUBLICATION DATA
Shaw, Bud, author.
Last night in the OR : a transplant surgeon's odyssey / Bud Shaw.
p. ; cm.
ISBN 978-0-14-751533-9
I. Title.
[DNLM: 1. Shaw, Bud. 2. Surgeons—Personal Narratives. 3. Transplantation—Personal Narratives.
WO 660]
RD27.35.S465
617.092—dc23
[B]
2015015828

Printed in the United States of America
1   3   5   7   9   10   8   6   4   2

Set in Ehrhardt MT Std
Designed by Leonard Telesca

*To Mom for showing me I could, to Dad for insisting I must, and to Frank, Shun, and Tom for showing me how*

# ACKNOWLEDGMENTS

We can't know, let alone remember all the people whose presence shapes what we become. At least I can't, and that's my excuse for failing to recognize all of those I'll now leave out, from the teachers who wouldn't let me get away with mediocrity and the guys who plucked my nineteen-year-old hitchhiking self from the freeway one frigid midnight and spent the next eighty miles trying to convince me to become gay, to the few teachers who worked so diligently at making students feel small and useless and the many more whose patience and faith often felt undeserved yet no less inspiring. They all played a role. That said, I want to recognize a few, not necessarily because they are most important, but because of their persistence.

When I was six years old, my mother made writing stories a work of joy and pride. I discovered an alternate world, one in which I had complete control. Her untimely death changed all worlds forever, infusing a reality I spent decades denying, no matter how real it became. My father taught me to be my own worst critic, from a similarly early age, less with his words than with the intangible signs of his joy and disappointment. He was also my hero, in so many wonderful ways. I miss him horribly.

Until Mr. Leonard Gwizdowski gave me my first C grade in fifth grade, I'd never received less than an A on anything. My mother was outraged with him, but Gwizzy stood his ground and I had to learn to study—really study. Six years later, Mrs. Mildred Veler gave me a required reading list separate from the other students'. "You're lazy," she said, and told me to start with Joyce. At Kenyon College, William Klein called my freshman prose verbose and obtuse, Galbraith Crump brought Shakespeare to life, Perry Lenz left America's great literary heritage deeply imprinted on my soul, and John Ward showed me unexpected joy in Smollett, Defoe, Bronte, Austen, Richardson, and Thackeray.

In Utah, so many surgeons proved critical to my training, including fellow residents, faculty, and dozens of private surgeons. I'm compelled to thank Frank Moody for setting such an annoyingly high standard for all of us and for always pushing me to be my best, despite my resistance. Gary Maxwell did more to lead me into transplantation than anyone. He inspired me with both his compassion and his astonishing abilities.

At Pittsburgh, I learned kidney transplantation from Tom Hakala, Tom Rosenthal, and Rod Taylor. I witnessed the unflinching integrity of Hank Bahnson, the unbreakable loyalty of Shun Iwatsuki, and the unstoppable drive of Tom Starzl. I owe more than I can ever express to Shun for being there to save my sorry ass over and over, and to Tom for being the font from which so many of my opportunities flowed.

In 1985, Bob Baker, Charlie Andrews, Bob Waldman, and, most important, Mike Sorrell and Bing Rikkers put together a proposal that pulled Bob Duckworth, Laurie Williams, Pat Wood, and me from Pittsburgh to Omaha, Nebraska. Together with Joe Anderson, Jim Chapin, Barb Hurlbert, Rod Markin, and Reed Peters, we forever changed the University of Nebraska Medical Center. My appreciation of the risks these people all accepted to make our work such a huge success is undying. Their expertise and dedication were indispensable.

I have many friends who read my earlier work and encouraged me to keep trying, including Jamie and Kyle, Carlos and Kathy, Bill and Chris, Steve and Genni, Mike Duff, and especially Dirk and Cath, whose friendship and faith are unflinching.

This book would never have come to life without the reboot I got from Steve Langan and the many participants in his Seven Doctors Project, the stubborn faith and encouragement of Amy Grace Loyd, and Jonis Agee's willingness to introduce me to Noah Ballard, who as my agent has brought wonders to my life. The confident enthusiasm that he and Matthew Daddona, the world's most gently tenacious editor, have brought to this work both astounds and delights me. Bo Caldwell and Ron Hansen unwittingly inspire me, most especially with their grace in success. Lee Gutkind introduced me to creative nonfiction during many sleepless nights together in Pittsburgh; his guidance has been invaluable.

My brother, Steve, and sisters, Mindy and Beth, may not agree with my versions of the events of our lives upon which I've shone so much wattage, but obviously my memory is better than theirs. I apologize for repetitively nagging them for details none of us see so clearly anymore, and I thank them for allowing me this indulgence.

I adored Carol for her art, her spirit, and her courage in facing death. I love Chris for our twenty-five years together, for our three children, and for loving me still. Ryan, Nat, and Joe are my real reason for existence, more for the joy they individually bring to the world than for notions of propagation.

Most of all, I thank Rebecca for resurrecting me, for nurturing my art, for keeping me honest, and for trusting me with her unfathomable love. I will always dance with you.

# AUTHOR'S NOTE

This is not a work of fiction. The events I have written about happened. That said, I must admit that in re-creating them, I have had to rely mostly on memory, not only mine but also that of many other people who were present or aware of what occurred. I was regularly surprised by how often my most indelible memories were not those of others. This sometimes led us to intense debate, and if we failed to resolve our confusion, I almost always stuck with my version because it felt most faithful to my experience.

Many of the stories involve patients. To protect their identities, I have changed or left out names, dates, places, and other details that are considered protected health information. These stories are thus merely representative of real experiences rather than the experiences of individual patients. I also found relevant aspects of some patients' stories in public records, including newspaper and television archives, obituaries, court records, and social media sites.

I was a transplant surgeon and most of my patients were recipients of liver transplants. Their stories are sacred. Their wait for a donor on the one hand and the heroism of organ donation on the other remain the most compelling part of the transplant story. I, almost as much as our patients, owe everything to the donors; more specifically, to their survivors who gave consent for donation. Without those acts of grace and courage, none of the patients I met would have survived, and I would never have experienced the joy and despair of trying to save them. Like many professionals who care for these patients, I lived every day of my clinical career with the horrible reality that more than half the people awaiting organs will never get them. We all long for the day when that is no longer true.

## PART ONE

# Expectations

# Pittsburgh 1981

## Initiation I

I was desperate to show how good I was that night. The patient was Max Stinson, a liver surgeon from Texas who, ironically, had a congenital form of liver disease that had finally progressed to liver failure. He was already opened up on the operating table when we got back from Virginia with the donor liver. Shun Iwatsuki was scrubbed in and had half a dozen other people helping him. Most of them left when we scrubbed in, but Shun stayed.

Dr. Starzl wasn't happy. Shun stood across from him. He had worked with Starzl in Denver and in the course of time, no one would be more important to my training. Dr. Hong from Shanghai stood to Starzl's right. His job was to retract the rib cage out of Starzl's way. He held on to the upper wound with both hands and leaned back like a water-skier. I would soon learn they called him the Human Retractor. Carlos Fernandez-Bueno was in his second year of training. He, too, had come from Denver, but by fall he would leave to accept a job he couldn't refuse at a prestigious East Coast center.

Starzl immediately began complaining. Shun kept silent and moved like a cat to retract something one way, push something else another, and, without a word, get Hong or Carlos to do something useful. I thought them telepathic, and this was their desperate attempt to appease the angry alpha. Already I doubted my own survival, useless as I was in this new world.

Shun's incision was shaped like the arms of a flattened Mercedes hood ornament, with a short vertical line coming downward from the base of the sternum to join a broad inverted V that stretched from one side of the abdomen to the other. Except for the donor in Virginia earlier that night, I'd never seen a body split so widely before.

I caught a glimpse of the liver lurking under the diaphragm. It was a shriveled, knobby, greenish-yellow lump. It was much too small for the space around it and it sloshed around in a puddle of blood every time the ventilator fired and pushed the diaphragm down.

Blood seemed to be coming from everywhere. The skin of the abdomen below the incision was a translucent, muddy yellow and coursed with giant blue veins running outward from the navel. I could see that the incision transected the course of some of these vessels, and when Starzl started removing sponges from around the edges, those veins rivered dark and red.

Starzl worked furiously to stop the bleeding. He lashed the open veins with silk sutures, grabbing needle holder after needle holder from the scrub tech as Carlos, then Hong, then Shun grabbed the ends and tied them down snug. I grabbed one when I saw my chance and broke it on the second throw.

"Shit," Starzl said and threw another stitch where I'd broken mine. This time I pulled too hard and the silk pulled out of the tissue and a blood torrent erupted. Carlos grabbed a sponge and pressed down, then got out of the way just as Starzl threw another stitch and tied it himself. He threw two more in the same spot, Carlos tied one and Shun the other, and finally the bleeding stopped. Shun frowned at me and vaguely shook his head.

I decided I could cut the ends of the stitches after they tied them. It was a job we gave to medical students. I was a trained surgeon, and a good one by all accounts back in Utah, where I'd finished my training less than a month earlier. So I grabbed a pair of scissors and cut the suture.

"Too short, goddammit," Starzl said. "That'll come loose and he'll bleed to death. Is that what you want?"

He laid down another stitch and Shun tied it, four throws, and I cut it.

"That's too long," Starzl whined. "Come on, now, Shun. Help me. He doesn't know what he's doing."


I worked my last day as a surgery resident at the University of Utah on June 30, 1981. I was thirty-one years old and I wanted to be a transplant surgeon. Dr. Thomas E. Starzl, the father of liver transplantation and perhaps the most renowned transplant surgeon in the world, had granted me a position to train with his team. He had moved his transplant program to Pittsburgh only six months earlier. I'd grown up in Ohio but I'd never been to Pittsburgh. I hated their football team, the Steelers, because they always beat my beloved Cleveland Browns.

My wife and I sold her car, gave the pickup truck to a friend, loaded my car and everything else into a rented truck, and headed for Ohio before dawn on Saturday, July 4. Eight miles east of

Point of Rocks, Wyoming, we hit an antelope yearling just as the sun rose above the rocks and sage. In Omaha we drove through smoke from a grass fire started by fireworks. In Ohio we unloaded the truck into my father's garage. I turned in the rental on Monday and called Starzl's Pittsburgh office the next day to ask advice about where to find a place to live.

I had arranged to take the month of July off. I figured five years of surgery residency and the specter of a two-year fellowship in transplant surgery justified it. We'd get a place to live and move in and I'd have nearly three weeks to do whatever I wanted. It would be like summer vacation when I was twelve.

The secretary in Starzl's office put me on hold. She said I needed to talk to someone else.

A female voice with a strange twang I'd later recognize as Pittsburghese came on. "Dr. Shaw? Where the hell are you?"

I said I was in Ohio, at my father's house.

"You were supposed to be here last week," she said. "July first, you know?"

She didn't understand. I'd talked to someone in March. A woman, I said. She promised to tell Dr. Starzl. I wasn't supposed to start until August.

I looked at the calendar on the wall above the phone in my dad's kitchen. "Saturday," I said. "August first."

"We don't have any information about that, Dr. Shaw," she said. "Dr. Starzl expected you last week. You've already missed two nights of call. I suggest you get your butt here today, tomorrow at the latest."

That was Tuesday morning. By Thursday night, we'd driven to Pittsburgh, put money down on a house a few blocks from the hospital, moved what we could with my dad's pickup, and found a grocery store open after ten p.m. I made my way to the hospital Friday morning and that evening my wife called to say the refrigerator had died. On Saturday I was on call when someone broke the window out of my car on the hill above the football stadium. They stole my toolbox and a Utah Jazz coffee mug. Sandee told me she wouldn't have parked up there.

Sandee was a nurse on the transplant team. She led me around the hospital my first day and told me where not to park on my second. When we went to the adult intensive care unit, a young woman in a short white coat and a shiny new stethoscope draped around her neck asked me to sign a petition banning liver transplants.

"They're unethical," she explained.

I wanted to laugh but she was so earnest. I said I was the new transplant fellow.

"Oh," she said. "Then you should definitely sign it."

She told me that in its first six months, the new transplant team had done six liver transplants and all six patients were dead.

"One time they ran out of blood across the whole state of Pennsylvania," she said. "Shut down surgery everywhere for days. It was awful. Gruesome, actually."

Sandee said the young woman was exaggerating. The blood shortage thing only happened one time; it was only in Allegheny County and only for part of a day.

"And the kids are all doing great," she said.

"The kids?" I said.

"All six of them," Sandee said. "Alive and well."

"Oh, so then you're batting five hundred?" The medical student stood with her head cocked, her clipboard resting on her hip.

"This ain't baseball, sweetie," Sandee said. She grabbed my arm and pulled me away and through the automatic door into the hall.

The University of Pittsburgh didn't have any place to put Starzl and his team when they arrived six months earlier, so they parked him in an unused laboratory, a huge room cleaved by long lab benches, complete with working sinks and gas jets. I shared a desk with the other transplant fellow. It was against the wall between two benches and the phone lines ran overhead, held to ceiling tile frames with bread-bag twist ties. Sandee said it was temporary.

"For how long?" I asked.

"Well, for at least the past seven months," she said.

I thought about the petition and wondered where we'd be in another seven months.

In my first meeting with Dr. Starzl I was asked about Dr. Moody. Moody was the chairman at Utah, where I'd trained. He had convinced me I should train with Starzl and he'd apparently written a flattering letter of recommendation.

"Frank Moody said you're a pretty good surgeon," Starzl said. "Do you think that's true?"

He looked at me and I saw something in his eyes that made me nervous. His question felt like a test.

"I guess so," I said. I felt like a thirteen-year-old. Then I told him I didn't expect to do liver transplants. It felt like a confession.

"Why not?" he asked.

I said I was interested in kidneys and pancreases.

"I see," he said. An odd frown came over his face, the space between his eyebrows pinching together. He turned in his chair, placed his hands on his knees, and sat for a moment looking at the wall; I could see the gentle slope of his shoulders and flawless profile of his nose. Then he stood up.

"Well, let's see," he said. He seemed uncertain what to say next as he looked almost furtively at the floor, then at me, then away.

I said I'd better be going and he smiled and held out his hand and we shook.

"Good to meet you, then," he said.

I nodded and left.

I'm not sure how I expected him to respond when I said I had no ambition to do liver transplants. I don't think I was seeking reassurance. Even now, I wonder if he thought I simply shared the mainstream judgment that liver transplantation was fruitless folly. When they learned of my decision to train under Starzl, some of the faculty in Utah and most of my friends thought I was making a huge mistake. One of the senior surgeons at a private hospital, a Mormon who'd been a transplant surgeon in New York, told me that liver transplantation was "just an expensive way to torture hopeless people before they die."

I wanted to laugh at them, call their skepticism ignorance, but I couldn't escape the doubts that still nagged me. Everything I knew about Starzl and liver transplantation was founded on years of rumored disasters coming out of his program in Colorado. By the time I had to decide where to seek training in organ transplant surgery, it seemed to me that things had changed.

Starzl's group had already begun working with a new drug that was much better at keeping the organ recipient's immune system from attacking the transplanted organ. It was called cyclosporine A, it promised to revolutionize the field, and Starzl was the only surgeon in the United States with access to it for use in patients getting either kidney or liver transplants. And then there was Frank Moody. Frank told me I needed to think big, that I should work and train with the best. Dr. Moody made me feel like going anywhere else would be a sellout, that working on the real frontier of transplantation was a far more glorious pursuit than playing it safe behind the front lines of that war.

"Come on now, take this!" Dr. Starzl had placed a retractor over the bowel.

Shun slapped my hand to get my attention and I grabbed the handle and held it exactly where he'd put it.

"Not like that," Starzl said. He ripped it out of my hand and repositioned it. "Like this."

I took it again and concentrated on not moving a twitch. I looked around the room and found it still so incomprehensible. In my prior experience, operations mostly involved a surgeon and a couple of assistants. Everyone knew who was in charge, who set the course, who was the decider. Starzl had five of us scrubbed in with him that night. The only other time I'd seen so many surgeons working on one case involved two people connected at the tops of their heads. It was in Utah and I was the resident on the plastic surgery team, and somehow I'd gotten a place there among the neurosurgeons and pediatric surgeons trying to separate these conjoined twins. It was, in my wide-eyed view, a clusterfuck. Every senior surgeon had an opinion about everything and I felt like a plaintiff's witness. I guess the difference with Starzl was that everyone knew that he, and only he, was in charge. My only role in his operating room was to stay out of everyone's way, particularly his, and I kept screwing that up.

Hong stood on a platform in order to be high enough, leaving about eighteen inches of space between him and Dr. Starzl. That was my space. Sometimes I'd try to help and lean in just a little and I'd bump Starzl's arm or his shoulder and he'd give me a hip check or an elbow to the chest. I didn't think it was intentional. Protecting his space was instinctual.

That first night, I couldn't yet see the pattern in anything we did. Often, Starzl just grabbed whatever hand was close by and shoved it where he wanted it, with no apparent regard for whose hand it was or what else it ought to do. "Shitfuckgoddamn," he'd hiss. "I can't see." "Don't hinder me, help me," he'd say when someone tried to help and failed. This was my initiation to the operating room of Dr. Thomas Starzl, and although I didn't know it at the time, these were but a few of the phrases I would learn to hate and mock and, in the distance of time and place, yearn to hear again.

# The Pros from Dover Fly to Virginia

## Initiation II

Earlier on that day when I met Dr. Stinson in the operating room, we were making rounds on the tenth floor with Shun when suddenly he stopped talking.

"Uh," he grunted. "Boss coming."

I turned and saw Dr. Starzl coming toward us down the long hallway. He looked energized. His steps were quick, with a bounce that suggested he wasn't so much in a hurry as he was excited. He was a thin, athletic man and looked much younger than his fifty-five years. An entourage trailed him. Two or three were Asian. They wore lab coats with sleeves rolled up to their wrists, the wide shoulders sagging toward their elbows and hems bouncing against their shins. Another was a tall, graying man in a tailored suit. He walked just behind Dr. Starzl and ahead of the Asians. He was someone important, but I'd never know his name. A young woman with red hair and a constant smile walked at Starzl's side, clutching a manila folder to her chest. As they came closer I saw Starzl had a lit cigarette pinched between his thumb and index finger. He carried it so that his hand sheltered the ember from the wind of his gait.

Shun seemed to know something. Starzl looked up briefly, their eyes met, and Shun nodded and took a deep breath. I'd only just met Shun. I hadn't figured out who he was yet.

"Well, let's see," Starzl said. "We've got an organ for Dr. Stinson."

I was among a group of surgery residents and nurses making rounds with Carlos and Shun. Starzl glanced quickly at us. His head movements were jerky, as though he were stealing a look and didn't want us to notice.

"Tonight?" Shun said.

Starzl looked down at his cigarette, then tucked his head under his shoulder and took a drag and blew it out quickly. We watched in silence.

"Well, let's see, then," he said.

He looked at Shun again, almost as though he'd been thinking about it; he looked at me and his frown vanished and he seemed to nod at me, almost nervously. I looked behind me to see if I'd been mistaken.

"Yes, of course tonight," he said. "Helicopter's picking us up . . . when?" He turned to the woman with the folder and bright red hair.

She looked at her watch. "Forty-five minutes," she said.

"It's a perfect liver," he said. "I need a first-class team, so let's see. . . ."

He looked us over and I tried to make eye contact.

"We can take . . . how many?"

"In addition to you and me?" she said. "Four."

"Right," he said.

Suddenly he jumped and dropped his cigarette to the floor, then looked at his palm. "Shitfuckgoddamn," he said and shook his hand. He crushed the cigarette out against the linoleum with the toe of his white Reebok.

"So, let's see. . . ."

He looked us over again and pointed to Carlos, the other transplant surgery fellow. "You," he said. Then he pointed to two of the Asians, a stocky, balding man and a young woman with hair to the collar of her white coat. "Hong and Wu?" he said and they both nodded. He looked at me and started to point but then stopped. "Well, I'd take you but you aren't interested in livers."

I protested. "That wasn't what I meant," I said. "What I meant was—"

"So you are interested."

"Yes," I said. "Very much."

"Well, all right. That's different," he said. He looked at Shun. "We'll call when we land."

He turned and bumped into the tall man in the suit, mumbled something, and walked off. The others hurried after him, except for the woman with the folder.

She told Carlos to be in the ER in twenty minutes. She looked at me and stuck out her hand. "I'm Mary Ann," she said. "You must be the new fellow."

I said I was.

"Twenty minutes," she said and walked after the team.

We'd fly at night, low and fast in a helicopter over cities and farms and the Allegheny Mountains. I was hoping to snag a seat up front.

A van took the five of us to the top of the hill behind the football stadium, where the helicopter waited. Dr. Starzl carried a wadded bright blue sleeping bag. He went to step into the helicopter and tripped over the end dragging along in the grass. The blades had begun to cross over our heads and a whistling came from the turbines when he reeled the bag in and pushed his way into the cabin. He and his blue bag sprawled across the back bench and Hong and Wu slid into the rear-facing bench and sat staring at him. He reached into a pocket in his cargo pants and pulled out an eye mask. He noticed Hong.

"Purloined from your airline," he said, grinning. "Air China."

Carlos was a big man and he hesitated before climbing past Starzl and wedging into the corner of the bench. Mary Ann climbed in on the other side of Starzl and I saw that there were no more seats. Elated, I turned toward the front door and bumped into a man wearing headphones coming out.

"We can only take five," he yelled.

Mary Ann looked at Dr. Starzl. He'd covered his head with the bag and seemed to be leaning against Carlos. She told Hong to stay behind.

Dr. Hong began to blink.

"Hong," she said and smiled at him, taking care to mouth her words slowly, in distinct syllables. "You. Need. To. Get. Out."

Dr. Wu was closest to the door; she got up, jumped out of the helicopter, and walked away from us without looking back. As I took her place, I looked after her and noticed the van was already gone. The copilot checked our seat belts and started to pull open the sleeping bag to check Starzl's, when Mary Ann grabbed his hand and shook her head. He hesitated, then nodded and closed the door, locking us in. As we rose up from the practice field I could see Dr. Wu making her way in the dark down Robinson Road and past the place where they'd broken into my car and stolen my toolbox.

I couldn't see much from my seat in the cabin. Hong had the window and he leaned his head against it and slept. We seemed always to be flying above a layer of broken clouds and without a moon. We landed on top of the large metropolitan hospital and walked down a flight of stairs and into an elevator.

I suppose it was all so glamorous—the helicopter arrival, the armed-guard escort to the locker room, the Trapper-John-in-Japan arrogance when we walked into the operating room like the pros from Dover—but I was too distracted to enjoy it. Or anxious; mostly I worried about being disliked for our hubris.

What unfolded over the next several hours was startling. I realize only now that in all those summers in the operating room with my dad, with other surgeons during my medical school rotations, and in five years as a resident working with surgeons in four private hospitals, a VA medical center, and a progressive university hospital, I'd never felt so much chaos or seen such licking aggression between two surgeons.

The encounter in Virginia was the first for the chief of surgery with Starzl and his team. In those days, before we published a paper proving otherwise, the running rumor among kidney transplant surgeons said that when taking out a liver, the Pittsburgh team damaged the donor kidneys. Maybe more challenging, surgeons, especially transplant surgeons of the era, considered the operating room their personal domain. They weren't accustomed to coordinating what goes on in there with other teams led by other alpha males.

Starzl understood all that. He knew he had to cultivate these relationships if he expected to get any donor livers from outside the Pittsburgh fiefdom. He tried to be affable, to make sure things went smoothly, but he also knew he had little margin for error in taking a liver. Maybe a kidney recipient could go back on dialysis if the new kidney didn't work, but people with failing livers had no such option.

Having a donor provide a liver or heart instead of just kidneys was a newsworthy event in those early days, one that brought out TV station and newspaper reporters to interview us on many of our donor runs. In Augusta, Georgia, later that year, three police cruisers met us at the airport and we drove 120 miles per hour to the hospital with trailing TV trucks unable to keep pace. In Cape Girardeau, Missouri, off-duty nurses brought in boiled shrimp, fried chicken, and snickerdoodles in case we were hungry, and in Dayton, Ohio, the mayor stopped by for a photo op. We were, after all, the pros from Dover.

The chief was helping the resident do the case and Starzl had little patience for the pace. Once gowned and gloved, we pushed our way up to the operating table, the resident backing away, still holding his forceps and scissors.

"You've done a fabulous job here," Starzl said. "Terrific, really. Mind if I do a little work up here?"

Hong tried to worm his way up to the other side of the table but the chief was an immovable green iceberg, so Hong stood back with his hands held close in front of him and rocked from one foot to the other. Carlos got in next to the chief and I stood back near the head of the table and waited for instructions. I could see the donor's neck and face and I realized he and I were about the same age.

The anesthesiologist saw me looking and arched his eyebrows. I moved closer to him on his stool and bent down.

"How'd he die?" I whispered.

"Fell out a second-story window," he said.

I frowned.

"Celebrating his promotion to manager."

I nodded and backed away.

Dr. Starzl pulled back the drapes to expose the chest. "Shit," he said. "We'll need to prep this."

The chief suggested we hold off for a while. "Twenty, thirty more minutes, Tom," he said. "So Anthony and I can get the kidneys ready."

The chief motioned for the resident to come back to the table and I moved out of the way to let him in.

Starzl asked the scrub tech for Betadine, the iodinated stuff we use to sterilize the skin before surgery, and he painted the chest and upper abdomen with an energy that scattered brown splotches over Hong's mask and gown, his eyelashes and neck.

"You go ahead down there," Starzl said. "We'll just do a little work here so we don't hold you up."

Starzl took a knife and extended the chief's midline abdominal incision all the way to the sternal notch, at the base of the neck. Then Carlos took a large hook knife and a hammer and split the sternum from top to bottom, while Hong and I tried to stay out of everyone's way. They put in the retractors to open the sternum wide and suddenly there was the dead man's heart jumping up and down in time with the beeping monitor, and just below it, the liver glistening and pink and more exposed than I'd ever seen it before. Starzl got Hong into his waterskiing position on one side of the rib cage and had me pull on the other while Carlos and he began moving the guts out of the way so they could get to the liver's blood vessels.

I could see this wasn't going to make the chief happy. Starzl needed the guts pushed downward toward the feet, and the chief needed them moved up toward the liver to allow him to see the kidneys and the lower part of the aorta and the vena cava.

"Why don't you take a short break now, Tom?"

"You've done a terrific job there . . . Tony, right?"

"Anthony," the resident said.

"Well, Anthony, you'll make a great surgeon someday."

"Seriously, Tom, give us thirty minutes and we'll be out of your way."

"Just let me get a look down here, make sure we haven't got unusual anatomy. A few minutes more, that's all."

I saw my chance to help and put a sponge on top of the stomach to pull it aside.

"Shit, now, help me, don't hinder me. Carlos, for God's sake, show him what to do. He doesn't know what he's doing."

He was talking about me, of course.

We left two hours later with what Starzl said was a perfect liver. The chief and Anthony were still working on getting the kidneys out when I took off my gown.

"Nice to meet you all," I said. I'm not sure they heard me.

# Max Stinson Gets a New Liver

## Initiation III

After what seemed like hundreds of sutures, the bleeding inside Dr. Stinson came under control. I felt suddenly exhausted. It was two or three in the morning by then and I was desperate to be of some use and barely able to stay awake, and then Starzl started working on the ruined liver. Every millimeter of progress took an eternity. We had to stitch and tie everything. I was thinking I'd have taken the electrocautery and just burned most of that tissue, been done with it. Then one of the ties came off and bleeding erupted and I thought maybe these veins were bigger prey than cautery could handle.

This was different surgery, different from anything I'd been trained to do. These people seemed to have lots of rules, even about things I took for granted. Like rules for tying off blood vessels before cutting them. Always four knots, not three. Always either 4-0 (thinner) or 2-0 (thicker) sizes of ties, never 3-0 (in between). Asking for 3-0 broke the rules. Not only did a request for 3-0 indicate indecision; it was also plain wrong. I was stupid for not knowing that the job of the moment required either 4-0 or 2-0 and not something in between. Mama Bear's choices made me a stupid surgeon. And they had rules for instruments. Only certain clamps were correct. Hand Dr. Starzl a hemostat with a subtle difference in shape from the one he wanted and you would break the rules. The wrong clamp went to the floor, or with a terrible clang into a stainless steel kick bucket.

Learning the rules was easy. That stuff never changed. But learning the complex steps of the operation felt impossible. I was anxious to be the best assistant Dr. Starzl had ever had, but just when I thought I had it figured out, he did something different. Sometimes it resulted in astonishing maneuvers and I felt lucky to witness his creativity. Too often it seemed like a distraction. I began to be suspicious. I thought that aside from the trailblazing gig, he also hated being predictable.

I had never seen anyone work on a liver so knobby and shrunken. The surgeons who had taught me always avoided working above and behind the liver, as though it were a no-man's-land where only trouble lay. Starzl destroyed that myth. It reminded me of the video of a Korean psychic surgeon on *60 Minutes*, the one in which the guy shoves his hand right through the patient's clothes and skin and swirls it around inside while making weird sounds and suddenly pulls his hand out holding a wriggling, ugly thing. Pure magic, just like the way Starzl freed up this horrible-looking wart of a liver and had it ready to remove before I even knew what he was doing.

Shun said something to Carlos and Carlos left the table. I thought he was taking a break but a few minutes later a sharp bang made me jump and I saw Carlos behind the back table beating on a towel with a steel mallet. He unfolded it and inside was a shattered plastic bag and a pile of crushed ice. He poured the ice into a steel basin, grabbed another bag of frozen IV fluid, wrapped it in a blue towel, and started banging away. And there was Mary Ann. She pushed a plastic spout into the end of a bag of liquid and began streaming it into the basin with the ice.

"Call Mary Ann," Starzl said. "We need the liver."

"I'm right here, Tom." Mary Ann lifted the Playmate cooler onto a stand and opened it up. She cut the strings with a scalpel blade and opened the first two bags and Carlos reached in and took the third. He laid it in the basin of slush and opened it.

"Oh," Starzl said after a few minutes. He looked up and saw Carlos inspecting the new liver. "Good," he said.

When we took the liver out of the belly of the donor earlier that night, I saw the vast emptiness left behind and felt awed but unconcerned. The donor was dead. It was a cadaver. I'd seen lots of things removed from cadavers in the lab or the morgue. That's how we learned anatomy all those years ago.

When Tom Starzl took the liver out of Dr. Stinson, he lifted it out by the clamp he'd put across the vena cava, like pulling a goblin's head out by the neck. He dropped it dripping black into a basin. I watched Shun hand it off to the tech, who laid it on the back table. I looked at the empty grotto where the liver had once lived and saw nothing but an impossible situation. I had a hard time comprehending that we were going to fix this horrendous problem, this unimaginable absence.

When Starzl called for the new liver, Carlos brought it up in a bag of icy water and Tom reached in, hauled it out like a tuna, and laid it on a towel. Fog rolled off the pale, glistening surface and Shun held it between his hands so that the opening of the vena cava at the top was exposed. Starzl put in some stay sutures, then lowered the liver into place while Shun and Carlos pulled up on the stays.

"Go, for God's sake," Starzl said. He tied the stay suture on his side and Shun tied his.

And then he sewed the new liver in and I couldn't believe what I was watching. I've done it myself a thousand times since then, but that night it struck me as the pinnacle of surgery, connecting all of these blood vessels and racing against time because once out of the ice, the liver was gradually warming up and we had to get the blood flowing to it in less than forty minutes. I tried to help by cutting the last suture, the one for the portal vein, but I fucked it up instead and cut it too short and the whole thing unraveled and Starzl had to do that vessel all over again and Jesus fucking Christ didn't Frank Moody teach me anything during those five years in Utah? Shun shook his head at me, squinted, and grumbled a warning every time I started to move. I felt my legs begin to shake, but all I could do was breathe more slowly and hope I didn't pass out. Then the suturing was done and Shun looked over the top of his glasses at the anesthesiologist.

"Are you ready?" Shun said.

The anesthesiologist jumped to his feet and turned up the flow rates of the intravenous fluids, his eyes darting back and forth over the various tracings on his monitor.

"Yes," he said. "I think we're ready."

"Come on, now," Starzl said. "Pay attention. You must be certain!"

Starzl reached up over the top of the liver, pulled it down with his left hand, and grabbed the big clamp on the upper vena cava with his right.

"OK?" he said.

He glanced at the anesthesiologist and before the guy could respond, he unclamped the upper vena cava. Then he took off the lower clamp and the liver looked like it was growing a purple birthmark as parts of it began to fill in with the venous blood. He waited, watching for any major bleeding. Seeing none, he removed the clamp from the portal vein. The color of the liver filled in rapidly, the birthmark giving way to a brighter hue that spread rapidly over the surface until the entire organ took on a brownish-red tone. He moved the liver around, looking on top and behind and underneath, shoved small sponges in here and there, dropped the liver into place, and stepped back from the table. I wanted to breathe again but I wasn't sure I could yet.

"Shun," he said.

"You want to do the artery?" Shun said.

"You have some work to do here first," Starzl said. "I want it bone-dry."

I moved out of the way to let Dr. Starzl out and by the time Shun got out from his side and came over beside me, Starzl was out the door. I took a deep breath and felt the squish of gelling blood in my shoes.

"He goes for smoke," Shun said in his accent. "Dry bone. What's he talking about? This is already dry bone."

"You want to take a break?" Carlos said to Shun. "I'll stay and work on this for a while."

"You go first," Shun said. "I'll go smoke when he comes back." He looked at me. "You want a break?"

I shook my head.

"Good. Go to the other side," he said.

With Hong holding the liver wherever Shun put it, we worked on getting the sponges out and finding a few bleeders here and there and stitching them up, until Shun looked up and asked for "many sponges" and filled in all the spaces around the liver with them.

"Dry as a bone," I said.

Shun grunted. "We wait," he said. He asked for a stool and sat with his head bent and his eyes closed.

Hong relaxed, working his fingers open and closed and shaking them vigorously.

"Cramps?" I said.

He looked at me and laughed. He set his hands on the drapes and blinked.

"Lot of pulling," I said, imitating how he'd been pulling with both hands on the rib cage.

He laughed again and started shifting his weight back and forth from one leg to the other.

"Where's home?" I said.

He blinked a few times, then reached in, lifted up the corner of the lap sponge, and looked at the liver. He dropped it and then tapped the liver gently with his finger.

"Good liver," he said and laughed. "Today good liver."

I looked at the monitors and worried that the heart rate was a little fast. I squeezed my eyelids together and felt the sting, then blinked a few times and looked around for another stool, but there were none. I felt like a sack of rags and I just wanted to sleep.

I didn't know how much time had gone by. Half an hour, an hour, I had no idea, but Starzl came back. Shun bolted upright; his stool rolled across the floor and hit a metal bucket and Hong woke up, laughing. I opened my eyes, found the sucker tip, and cleared away a pool collecting in the trough of drapes at my waist. It was more blood than I expected.

"Well, let's see," Starzl said. He rammed his hand into the second glove and the rubber cuff snapped as he tripped over a kick bucket. Hong grabbed his arm.

"What are you doing, Hong?" Starzl jerked his arm out of Hong's grasp, kicked at the basin, and came to stand between Hong and me. He grabbed the sucker from me and at the same time ripped the darkening towels off the field and flung them aside, onto Hong's face and gown.

"Hong, you're contaminated," he said.

While Hong stepped away to get a new gown and gloves, Starzl looked around the liver and sucked out blood. I heard him say something to Shun, but Shun kept quiet and probed blindly with the sucker while staring at the top of Starzl's head bent low across the table.

"Hold this," Dr. Starzl said. "Come on, shit!"

He meant me. I reached in and retracted the bowel.

"Like this!" he said and pushed down hard on my hand, but from where I stood I couldn't pull in that direction.

"Shit, he can't do it. Shun! You'll have to help him."

I looked at Shun and he seemed immobilized by something—fatigue, perhaps.

"Shitfuckgoddamn," Starzl said. "I don't want anyone here who doesn't believe in life!"

Suddenly, Shun shifted into some sort of overdrive. In the months to come, I'd learn his frenetic movements were deliberate, meant to show he was serious, that he cared about life, that he was going to do everything possible to help. It also meant that he had seen something in Starzl's behavior that suggested trouble was just one wrong move away. I think Dr. Starzl always knew how he wanted things set up, but he either wouldn't or couldn't tell us. Instead, he carped about our lack of interest, our desire to hinder him, our inattention and neglect, our incompetence, until someone did the right thing. Hong knew when to duck to avoid his elbow but it was up to Shun to put things into place, and he did so deftly. Most of the time.

Carlos came back about then. He was able to hold the small bowel out of the way where I couldn't. Starzl put a small clamp on Dr. Stinson's artery.

"Wonder Glasses," he said.

"What?" the tech said.

"And call Dr. Wu," Starzl said. "I need the powers of Wu."

Carlos told the circulator to call Sandee. "Wu doesn't have a pager," he said. "Sandee may know where she is."

Shun told the tech Starzl wanted his loupes.

We passed the wait watching Starzl try over and over again to get the loupes adjusted. They were an older pair, not custom-made but adjustable for any face, with slide screws that moved the individual lenses up and down, in and out.

"Shit, Shun. Have you been using these?" Starzl said.

"I never use your Wonder Glasses," Shun said.

"Someone did," Starzl said. He tried to adjust them with a towel over his hand. "I had them perfect," he said.

"No one else uses them," Shun said in a low voice. "No one knows where to find those Wonder Glasses."

"Shit, don't debate me, Shun." Starzl threw the towel onto the floor and used his gloved hands to turn the screws and slide the lenses and push the bridge up and down along his nose.

The serene and powerful Dr. Wu arrived about then. Dr. Starzl stopped fiddling with the loupes and looked into the abdomen at the arteries.

Dr. Wu stood nearby, across from me, but all I could see was her head above the table. Her face was almost entirely covered by the paper mask. She stood motionless in the shadows. Starzl seemed to know she was there without looking.

"Dr. Wu," he said. "Please help me."

Wu nodded and disappeared. Starzl told me to hold something and Carlos poked around in the abdomen with a sucker. I saw Wu reappear a few moments later, gowned and gloved and rising up like a rock star between Shun and Carlos.

"I'll take a break," Shun said.

"No, Shun. Not now."

Shun had stepped off the stand. He stopped with his hands folded. "Dr. Wu's here," he said.

"What?" Starzl looked up. "Oh. Wu. Good."

"So, take a break now?"

"What? Oh, yes. Go ahead, Shun."

Shun left and with Wu opposite him and holding the ends of the suture out of the way, Dr. Starzl sewed together the ends of the two arteries. His movements were more fluid than I'd seen all night and in the new quiet I heard his calm breathing. When he released the clamps, the pulse was bounding. He fixed a small leak with a single stitch, then stood up and inspected the surface of the liver. He rubbed it with his hands and then lifted it and looked behind and under it, then let it drop back in and rubbed it again.

"It's a perfect liver," he said.

With fully oxygenated blood flowing into it, a brighter pink filled in the remaining dark patches on the surface.

"Look at that," he said.

With that, he left. Wu disappeared. Carlos and Hong and I looked at each other.

"Why don't you take a break now?" Carlos said.

Hong leaned into me and when I looked at him he lifted his elbow to point away from the table and nodded.

"That's OK," I said. "Why don't you go?"

"I won't be long," Carlos said.

Hong and I managed to talk. I got that he was from Shanghai, Wu was from Nanjing, and there was someone named Ming from Wuhan. Hong said he knew about Utah.

"Joseph Smith," he said.

"Brigham Young," I said and laughed. "Church of the Latter Day Saints."

Hong tilted his head.

"Mormons," I said.

"Yes!" he said and slapped the top of my hand.

I asked him about Wu.

"She from Nanjing," he said.

"But why did Starzl ask her to come?"

Hong tilted his head and rocked back and forth a bit more.

"It's like he thinks she has some sort of mystical powers," I said.

"Dr. Wu," Hong said. "She from Nanjing. I from Shanghai."

Shun came back. He told me to take a break and I did. I went to the lounge and drank some tepid coffee from an urn. It was the color of mud and I realized it already had cream or creamer or maybe milk mixed in. I was alone and the more I drank, the more exhausted I felt.

I threw away my paper cup and got up and wandered through the halls behind the operating suites. Someone had left a gurney outside our room. It was probably for Dr. Stinson. I hopped up onto it and lay down, for just a minute.


A nurse shook me awake. She said she needed the gurney. I thought I'd slept for twenty minutes, maybe thirty. I went back to the operating room and got as close as I could to see. I stood there staring at the open abdomen, trying to figure out what they were doing.

"Can I help you?"

A nurse had come up beside me. She was trying to see my face. It was a different nurse.

"I think I'll scrub back in," I said. I held up my hands. "Eight and a half, brown."

"I'm sorry," she said. "Who are you?"

I told her I was the new transplant fellow. I'd just come back from a little break, I said.

"Oh," she said. "You're a little late. This is Dr. Watson's gallbladder."

I looked at the clock. It was after ten a.m. I'd slept for more than four hours.

"You're welcome to watch."

A surgeon with gray hair and wire-rimmed glasses looked over the table at me. "You might learn something," he said. "Dr. Ladowski here has become a mighty fine surgeon."


Shun was smoking in the OR doctor's lounge when I went to change.

"You feel strong?" he said.

I poured myself a cup of mud and sat down in a chair beside him.

"I fell asleep," I said.

"Carlos said you went home to bed."

"I was on a gurney. In the back hall."

Shun took a drag and blew half a dozen smoke rings over his head. "Starzl says you're weak," he said.

I felt nauseous. Shun held his Marlboro between his index and middle fingers with his wrist cocked back. The coffee smelled of cigarette ash.

"He says you're not trained well."

I didn't know what to say. Shun blew smoke above his head again and recrossed his legs.

"Did I miss anything?" I said.

He never blinked. "Never take a break unless he takes a break and always come back before him. No rest until patient is dry bone."

I told him I didn't need any breaks.

He grunted. "Don't be crazy."

I left Shun in the lounge and found Dr. Stinson in the ICU.

"Are you the new fellow?" Dr. Stinson's nurse wore a pink smock with dolphins swimming on it. She handed me a loose-leaf notebook with "Stinson" written on the spine.

"We need orders," she said. "The intern said to page you."

I took the chart from her and opened it to the order pages. They were all blank.

"Where have you been?" she said.

She was talking to me.

"I don't have a pager yet."

"I see," she said.

We stared at each other for a moment.

"Orders?" she said.

I nodded and pulled out a stool from under the counter and sat down. I reached for a pen but my pocket was empty. She pulled a pen out of the pocket of her smock, clicked it, and handed it to me. "Lose this and I'll come after you in your sleep."

I took the chart and went into Dr. Stinson's room. He was on a ventilator. The dressing on his wound looked dry and his belly was flat. There was a full column of clear yellow urine and the bag at the end was half-full. I took his hand and leaned over and asked him to squeeze my hand. I didn't expect him to respond. I didn't even expect him to be alive after what I'd seen, but I felt a twitch. I asked him again and that time it was stronger.

"Your transplant's all done," I said, close to his ear. "You're going to be fine."

# A Surgeon Is Someone
# Who Can Fix People

I remember awakening after midnight to the sound of my father's whistle and his footsteps on the wooden floor when he came home after a call from the hospital, like the one about a motorcycle crash and a leg nearly cut clean off. I always fell back asleep certain he'd fixed everything.

He had to fix me more than once. Before I learned to walk, I drove a stroller off the front porch and split my forehead on the sidewalk. He sewed it up and my mother's bridge club friends judged the job perfect. Three days later, I rolled the cart through a screen door and ripped the perfect closure open. To this day, the scar runs down my forehead at an angle.

By my first year in college, my father had sewn me up six or seven more times. Four more wounds involved my face or head; the rest, hands and feet.

The winter I was nine I split my cheek on a bed rail playing Swamp Fox. That was our favorite Disney TV show that year, or month, and I was spending the night with my best friend, Jim. I was the Fox and Jim the vile redcoat, Colonel Tarlton.

As was customary, the colonel chased the Fox through the swamp. The Fox dove for cover under the Spanish moss draping an ancient cypress tree disguised as Jim's lower bunk. As luck would have it, and as is a danger in the swamp, the carpet of ground cover slid out from underfoot and my dive came up short and suddenly the Fox was bleeding.

Mrs. Tatman sat me down at the kitchen table and dabbed my cheek with a cold dishrag. Mr. Tatman looked over her shoulder.

"Ooh, that's nasty," he said. He worked on the line at Pennington Bread and wore his red-and-white work shirt at home.

Mrs. Tatman said it might just need some butterflies.

"Elmer," she said. "Call Dr. Shaw."

No one ever came in the Tatman front door, but my father did. He wore his winter hat, the one with the small brim and earflaps that he unfolded from inside when it was really cold. He took one look and said I'd need a couple of sutures.

In the car he was quiet and I sat wishing he'd say something, like how they have this new way to numb a cut that doesn't hurt at all and that they'd figured out how to remove that scary chemical smell of the place, and that the really bright lights—the ones that blind you when you're lying there staring at the ceiling trying not to move while every voice in your head tells you to jump up and run for the door—well, those lights aren't working this week. When we pulled up to the hospital and I saw the red neon that read EMERGENCY I figured he'd see my cut under better light and decide on a couple of butterflies. One of the kids at school got those and it didn't involve needles.

As I lay on the operating table, my face half-covered with green towels and the nurse telling me stories about her cat, I heard my father whistling in the hallway and the water running and I knew he was scrubbing his hands. His voice was close when he said thank you to the nurse and I heard the snap of the rubber gloves and then the stool with the metal wheels rolling toward me. I wanted to watch but I only had a small window through a gap in the towels. I heard metal clicks and sharp clanks and I knew he'd be getting ready to use the glass syringe with the long silver needle on the end.

"OK, now," he said right beside me. "You'll feel a bee sting."

It wasn't the needle that hurt so much as the numbing medicine, and it burned worse than any bee sting I could recall.

Then it stopped hurting and I felt him pushing on my face and it was like the skin wasn't connected to me anymore. I felt him stitching, the pushing down on my face as the needle went in, then an eerie tug on my facial skin as he dragged the thread through. Through my window, I could see part of his face. His metal-rimmed glasses flashed in the light. At the top of his pull, I caught a glimpse of the curved needle, gripped in the jaws of the silvery clamp, his brown-gloved fingers working the ringed handles. Lying on my back, a green cloth partially covering my face, the breath of disinfectant everywhere, his voice close by but detached, I felt safe.

Such power, to make someone feel safe in the face of something so scary. I look back now and wonder if I had a sense of that power then.

# A Hero in Ohio

My father was a surgeon, a really good one by all accounts, including mine once I knew what this constituted. We lived in a town of some twelve thousand in the flattest county of south-central Ohio and by the time I was ten years old I realized he was the town hero. He could cut people open and remove stuff they were better off without or sew a finger back on if the corn picker hadn't mangled it too badly, or remove a foot when it had. I saw it every time we went out to dinner at the J&J Drive-in or Anderson's Diner. People couldn't leave him alone. It didn't matter we were trying to eat. Some guy'd come over and pull up his shirt-tail to show us a belly scar, or an old lady would put a bear hug on him and say the word *uterus*, or a tall man in a cowboy hat would flex and stretch his arm and say, "Better than new, Doc."

I didn't grow up wanting to become a surgeon. A few months after my tenth birthday I switched from fireman to marine biologist after snorkeling with manatees in Crystal River, Florida. After that I wanted to spend life with an Aqua-Lung strapped to my back and a view of everything underwater. I was eleven and had yet to learn to scuba dive, but already I was terrified of getting the bends, so I memorized the decompression tables. My science projects were mostly about the invention of the Aqua-Lung and how scuba regulators worked. I knew all about Jacques Cousteau and that in 1960 he told a reporter for *Time* magazine,

"Under water, man becomes an archangel." I got access to diving gear catalogs and began saving the money I made mowing lawns, raking leaves, and shoveling snow to buy a weight belt, a thirty-eight-cubic-foot scuba tank, a two-stage regulator, and a new pair of Jet Fins. Everyone insisted I would follow in my father's footsteps but I thought marine biology was a lot more fun.

For a long time, I didn't think my dad cared much about what I wanted to be, but one evening at the dinner table, he told me marine biologists didn't make much money. "And a good boat will cost you a lot of money," he said. That was before my mother got cancer, and after she died I didn't think much about being a marine biologist, and by then I'd stopped reading about skin diving and started worrying about how to get my arm around Patty's shoulders at the movies. By the time I flunked my first driver's test a few years later, James Bond had replaced Jacques Cousteau and all my earnings went to taking Patty to the Friday all-you-can-eat seafood buffet at Lincoln Lodge in Columbus.

I finally passed my driver's test, and I began considering the possibility that doctors were community frauds. Nothing my dad did made me feel that way. I just found the worshipping business gratuitous. Doctors were simply doing their jobs; they were paid to save people, to treat their pneumonias, their diabetes, their bunions and ingrown toenails. In the highly sophisticated view of the world that came with a driver's license, I thought they all seemed too ready to indulge all that veneration. I didn't want to take myself so seriously. My heroes were all so much more humble. My dad likely thought me sanctimonious, especially when a patient at Anderson's Diner asked me if I planned to follow in his footsteps. No way, I said, as though it were obvious.

# An Exclusive Intimacy

When my mother died I lost faith in God, doctors, and Dad. For whatever reason, none of them saved her. Years would pass before I recognized my father's devastation. In the meantime, I just felt abandoned. He remarried in eighteen months and although I understood the claims that Billie wasn't supposed to replace our mom, my father grew more distant. I decided this was just part of growing up, of becoming a man. Throughout high school, I didn't get much of a sense that Dad cared a whit what I might want to do when I grew up. Maybe if my mother had survived my preadolescence, someone would have been there to encourage me one way or the other. But Dad? Sometimes it felt like he was trying to push me away from a medical career. I was suspicious he was using a weird kind of reverse psychology on me.

I was seventeen before he agreed to let me drive his Thunderbird convertible. It was summer and I wanted to take Patty to the drive-in but I needed to wash and wax it, so I dropped by the hospital early Friday afternoon to exchange cars with him.

He was in the emergency room. A nurse took me to the room where he was working.

"You're just in time," he said. "Stand over here."

He was wearing a pale green surgical mask, a light blue surgical gown, and yellow rubber gloves. An elderly woman was lying prone on a table that was flexed in the middle so that her butt pointed obliquely upward. A slender silver tube poked out from under the green sheet that covered her lower torso. My father held on to the tube with one hand and pointed to where I should stand with the other. As I moved around to the spot, I could see where the tube disappeared into the woman's buttocks. She didn't seem particularly bothered, though. A nurse stood next to her, now and then whispering to the old woman.

"Can you see?' he asked. I nodded. I was feeling a little sick to my stomach.

"Here. Come a little closer."

I said I was fine.

"No, come closer," he said, "or you'll miss the best part."

He turned his attention back to the tube. He looked through a glass cover on the end of the tube and now and then gave a rubber bulb attached to the side a couple of squeezes to pump air into her bowel. He seemed to be pushing the tube farther and farther up the woman's butt, pausing whenever the woman groaned a little, then rocking it a bit and pushing it an inch or so more. He talked the entire time.

"Mrs. Smalley here lives over at Screnity Oaks and woke up this morning with belly pain and vomiting."

Twist, rock it side-to-side, push, a little groan.

"She's got a volvulus." He looked at me. "Twisted bowel. We're going to untwist it for her."

Finally he stopped. "You sure you can see OK? Because this is the best part."

I moved a little closer, my stomach creeping up into my

mouth. I was wearing a white T-shirt, cutoffs, and sandals. I had to bend over slightly to see under the sheet, hands on my knees. I was still about six feet away from the end of the tube.

"Here goes!" He untwisted a small screw on the side of the lens covering the end of the tube and flipped the glass open.

Brown liquid shot out of the tube and covered me from the waist down. The room filled with the foulest stench I had ever smelled. I jumped back and stood with my arms out. I gagged and covered my mouth, then noticed that my hands were dripping brown mush speckled yellow with kernels of sweet corn.

"Still want to be a doctor?" He laughed at first, but then said something to the nurse. She left and came back quickly with a couple of white towels, a bottle of reddish liquid, and a pair of green scrub pants. Disinfectant soap, she said. She led me over to a big sink and showed me how to turn the water on by pressing my foot on a pedal underneath. She pulled a curtain so I could change into the pants in privacy.

When I finished, Dad was gone and the patient was on a wheeled cart. The nurse came over to inspect me.

"You might want to throw those away," she said, pointing at my stained shorts on the floor. She started to push the cart down the hall then turned back.

"Oh yeah. He left the keys at the front desk."


At the drive-in movie, I pressed for *How to Steal a Million*, with Peter O'Toole, but my girlfriend chose *Born Free*. In the end, it didn't matter. I didn't tell her about what had happened at the hospital, but I couldn't get it out of my head. I thought it was gross, and I was still angry my dad was such an ass, but I also felt something else, a kind of thrill that I couldn't explain, an exclusive intimacy with the human body that felt like power.

# The First Notes of a Siren Song

During the summer following my graduation from college in 1972, Bobby Fischer defeated Boris Spassky in Reykjavík to become World Chess Champion, Jane Fonda visited North Vietnam, the last American ground troops were withdrawn from Vietnam, eight members of Black September killed eleven members of the Israeli Olympic team in Munich, where Mark Spitz won seven gold medals, Bernstein and Woodward broke the story of Watergate, Hurricane Agnes killed 117 on the East Coast, and health officials made public the details of the Tuskegee syphilis experiment. As consequential as those events were to the rest of the world, I paid little attention. I was twenty-two years old and I cared only about getting out of combining wheat, baling hay, castrating pigs, sealing asphalt, and any of the other jobs I'd had in summers past, and hospitals were air-conditioned and clean and the pay wasn't bad. So, four days before the Black Hills Flood killed 238, and nearly three months before I started medical school, I took a job as a scrub technician in the operating rooms of Fayette County Memorial Hospital, in Washington Court House, Ohio. I worked for my father and his partner, TJ, and as far as I was concerned, I still had no aspirations of becoming a surgeon.

Billie, my stepmother, taught me how to be a scrub nurse.

When I made a mistake putting on my gloves, invariably a mistake I hadn't noticed, she sent me back to the sink to scrub my hands all over again. If I dropped a hand half an inch below where she imagined my waistline to be under my billowing sterile gown, she reminded me that everything below my waist was considered contaminated and sent me back to the sink. If I handed TJ a long skinny hemostat when he asked for a short fat one, she rapped the back of my hand with a clamp and handed him what he needed. I'd like to say I later grew to appreciate her discipline, but I didn't. I would have been my harshest critic had she given me a chance. I went into the job terrified I couldn't take it—the sight of a scalpel slashing through skin with blood trailing after, the airy in-and-out sounds of the anesthesia machine, the sickening sweet smell of disinfectant. All of that was overshadowed by my fear of Billie. Maybe that's how I got through it all.

I had no basis for judging surgical skills, yet right away I noticed the difference between my dad and TJ. They were both good, but my father's spare movements and sheer speed enchanted me. That first summer I learned to keep up with TJ, but I always slowed my father down. His mind stayed three moves ahead of his hands and I was two steps behind everything. He'd have his hand out waiting for a needle holder while I tried to figure out which of the dozens of needle holders he might want and where it was.

I was better the second summer, and they promoted me to surgical assistant. I had a year of medical school in hand, a year I hated with all my soul, but I felt useful in the operating room even if nothing I'd learned in school was of any use there. Billie began leaving me alone. Being a surgical assistant freed me from having to manage the instruments, which must have been a relief to the surgeons, probably more so for Dad than TJ. Now I held retractors and cut sutures and otherwise stayed out of the way of the scrub nurse and the surgeon and the general practice doctors who liked to scrub in and play copilot on their patients and found me an enthusiastic pain in the ass.

# In Cleveland, Struggling to Belong

During the first two years of medical school I worked part-time with surgeons in community hospitals, sometimes as a paid employee, other times as a volunteer. I learned that some surgeons weren't very good and that some were terrifyingly bad. I fancied myself the well-trained, highly experienced surgical assistant and tried to help when I could, but most of the time they told me to keep my hands and opinions to myself. One scrub nurse told me I should consider a career in dentistry and the surgeon, maybe the worst I've ever seen, laughed.

"Maybe a pathologist," he said. "He'd always be too late to do any harm."

I worked the summer between my second and third years in medical school as a surgical assistant in a hospital in West Cleveland. The surgeons there were hardworking, efficient, very confident, and nearly as good as my dad. Some of them did more complicated surgery than Dad did and I saw the field opening up to a larger world of possibilities. I worked with a young assistant named Asan. He'd been a full-fledged surgeon in his country but he wanted to do a residency program in the U.S. "Oh, to be the best," he said. Dr. Asan taught me how to start IVs in patients without obvious veins, how to insert a catheter into a large vein under the collarbone and thread it into the heart without puncturing the lung or a nearby artery, how to set up a surgery case for each of the surgeons. Sometimes the surgeon poked his head in the door and told Asan to get started, that he'd be just a minute. Sometimes a minute or two became twenty and by the time the surgeon arrived we'd have the abdomen open and the retractors set up, and the surgeon would look around and smile and say, "Excellent work, Asan."

Asan explained everything he was doing and why, and one time he let me make the incision. I remembered my dad explaining how he drew the scalpel quickly across the skin so that the blade went straight down to the muscle, but I'd never tried it myself; when I did it for Asan I pushed a little too hard and the blade went through the skin, the underlying fat, and the tendinous juncture between the muscles. It stopped short of going clear through the last layer and into the abdomen, but Asan and I both gasped and I felt my heart pounding in my throat.

"Oh, God," Asan said. He looked at the scrub nurse, who looked down at her tray and shook her head.

"I'm so sorry," I said.

Asan grabbed a sponge and swabbed at the wound and we saw very little bleeding. The incision had gone right down the center and missed the muscles completely. We heard the sink gushing outside the door and knew the attending surgeon was scrubbing. Asan handed me a pair of forceps, his hands shaking a bit, and I knew he wanted to get the incision open and have the retractor in place by the time that door opened. And that's what we did, and when the attending walked up to the table and looked things over, he smiled at Asan behind his mask.

"Excellent work, Asan."

Asan invited me to his apartment for lunch that day and I had my first taste of curry and basmati rice. He took a drink of tea and sat back in his chair, then looked at me and shook his head.

"We were lucky today," he said.

I nodded and swallowed. "I know. Scared the shit out of me."

He told me what I'd done wrong, how it took a lot of practice to do what I'd tried to do without cutting too far.

"You could have sliced a big gash in the colon. Cut it in two," he said.

"Or the stomach, or the liver," I said.

"We'd both be fired," he said. "I cannot make myself get fired, you know?"

I nodded.

"I have plans and I am careful. I never try what I don't know works, what I don't know."

"I know. I—"

"You see, I have my wife and my little baby and we must not get fired or have anything go bad for us, so we make certain we are careful. Always. Surgeons must always be careful but I must be even more careful."

He stood up and put his plate in the sink.

"You like curry?" he said.

# A Failure in Psychiatry

During my third-year surgery rotation, I was always eager to show what I could do—tie knots better than the intern, arrange the retractors to get better exposure, cut sutures just the right length. I was rebuffed over and over. I came to understand that they'd let me join their little crew, but only if I knew my place at the bottom. I was frustrated not so much by not getting credit for having what I thought were useful skills, but because I'd seen many of these operations go much quicker and easier when my dad or TJ was at the helm. I didn't mind being a loyal deckhand as long as I had faith in the captain. I once told the chief resident after a three-hour gallbladder surgery that I'd seen my dad do that operation in twenty-nine minutes. X-rays and all, I said.

"You're so full of shit," he said.

Following the surgery rotation, I moved on to the required stint in psychiatry. I was stationed at the Cleveland Veterans Administration Hospital and didn't know what to expect. I'd never met a psychiatrist before. I wasn't really sure what they did, but I was curious and I thought maybe I could be more useful as a psychiatrist.

One of the first patients I was assigned was Ike, a black man in his early forties who told me he'd lost his "nature." We came to an understanding that he was having trouble getting an erection. I also figured out that Ike was depressed and a recidivistic alcoholic. We had a nice long chat. He didn't seem to think his drinking was a problem, insisting the fights were all started by others, that his girlfriends didn't understand him, and none of that was the point. All he wanted was his nature back.

"Let me see the pecker doctor," he said. "I know you got them here. Lots of vets got the same trouble and the dick doc's the one sets them straight. You know what I mean, Doc?"

"I think your dick won't get hard because you drink too much," I said.

He told me that was some fancy bullshit, that he'd never had anyone try that one on him before.

I gave him a schedule of AA meetings and told him I wanted to see him again early the next week to see how he was doing.

I presented his case to my supervisor and he said there wasn't any treatment worth trying until he agreed to stop drinking. He also said we had no reason to consult a urologist, that he thought Ike was also depressed, but that we couldn't give him any drugs for that.

"Not the way he drinks," he said.

That Sunday, the operator at the VA hospital put a call through to my house. We had friends over for dinner and I was trying to get the charcoal lit.

"Doc, you got to help me."

It was Ike. He said he'd had enough. "I come to the end of the road, Doc, know what I mean?"

I said I didn't.

"Well, let me put it this way. Listen to this, OK? Just listen."

I heard a click followed by a louder, metallic noise, like the hammer on my .22 revolver, only bigger.

"Know what that is?"

I lied, said I didn't. He told me to listen again and repeated it, this time pulling the trigger slowly so I could hear the click of the cylinder moving and the fall of the hammer.

"Now you know, don't you? Don't lie to me, Doc."

I asked him what he wanted. He said he was going to do it, that he was sick and tired of living like this, that a man wasn't worth shit if he didn't have his nature. Better to just end it now.

"Blow my worthless brains out," he said. "Know what I mean, Doc?"

"Give me your number, OK? I'm going to make a call and see if we can get you over to the VA, OK?"

Ike said the VA was bullshit. "They can't help me. You and them, you're all alike, if you catch my meaning."

I'd had no experience handling this kind of thing. I got his number, then put a call in to the VA operator and asked to talk to my supervisor. I held while she rang his home; he answered and I told him about Ike and said I was really worried and didn't know if I should call 911 or the cops or what but that maybe Ike should go to the hospital.

"How did you get my home number?" he said.

I told him I went through the operator and he said I had no business calling him at home on a Sunday. He told me Ike was a manipulative alcoholic and that he was too narcissistic to kill himself and that I should have just told him to keep his next clinic appointment, like we'd arranged. Then he hung up.

I tried to call Ike back but got no answer. He never showed up at the clinic again. I didn't see any reports in the paper about a man blowing his brains out.

I felt so stupid. I felt like I should have known better, that maybe my supervisor had taught us about this sort of thing and I'd dozed off again. I wanted a chance to redeem myself.

That chance came less than a week later. I was doing my night-call obligation at the VA, which usually meant doing half a dozen admissions. That meant only the history and physical at most hospitals, but not after-hours at the Cleveland VA hospital, where we were the nighttime lab and X-ray technicians as well as patient transport.

The sixth patient of the night came in about eight thirty. Mr. Mullen was a fifty-seven-year-old man with depression. He said that the visits to the clinic weren't working and that he had started thinking about suicide so someone told him to come in to be admitted to the hospital. I put him in a wheelchair and we went down the hall to the lab, where I drew his blood, had him pee into a cup, and did a blood count and a urine analysis, recording the results on his chart. Then it was off to radiology, where I asked his weight and used the chart taped to the door to figure out the settings for the X-ray machine. I had him stand up, shot two films, ran them through the developer machine, and stuck them on the view box while he watched.

Mr. Mullen never said a word.

Once I got him to his room I sat down to take his history and found out he had a twenty-year-old daughter and that he was sure she hated him. He said she was running around with boys, fucking lots of boys, and that when he tried to discipline her she ignored him. She said he was just a stupid old man and that he could go fuck himself because she was going to do what she wanted. I started my physical examination as he talked and when I got to his abdomen I felt a large mass right in the middle near his belly button. It was hard and pulsatile. I listened to it with my stethoscope and heard blood rushing through it more loudly than was normal. I checked his other pulses—neck, wrists, groin, and feet—and they seemed fine. I interrupted him going on about his daughter and asked him other questions related to symptoms of arteriosclerosis and heart disease and he denied all that.

I asked him if he had back pain.

"Yeah, this week it's a lot worse."

Did it radiate into his groin or his testicles?

"Yeah, mostly the left," he said. "Kind of a ache that seems like it's connected to my back pain—I can't explain that; it's just how I feel sometimes."

I told him we had to go back to radiology. I had him lie down on the table and shot a film across his abdomen from right to left. He had to hold the film plate for me and he kept shaking, so I handed him the button, went over to hold the film myself, and told him when to push the button. I took a wild guess on the settings and they were pretty far off, but the film showed me what I wanted. He had a large rim of calcified tissue bulging toward his belly button. I'd seen this the month before when I'd rotated onto vascular surgery for two weeks, and I knew it was what they called an eggshell sign, which meant he had an aneurysm of his aorta. When I measured it on the film, even subtracting 20 percent for the magnification that I thought was likely, it was still a huge one, and with the back pain radiating to his groin, I was worried it might be rupturing. I knew that happens to people once their aneurysms get above a certain size, and this one was way bigger than that.

I took him back to his room and told him I was worried about the large blood vessel in his belly, the artery called the aorta, which carries blood from the heart to the kidneys and then down to the legs. I said it had a big blister on the front and that it was dangerous because it could burst, and if it did, he had a lot worse chance of surviving than if the surgeons could fix it now.

"So you're saying I came in here because I talked about killing myself and now you've found something that could have done the job for me?"

I nodded.

"What a fucking world."

I called the chief resident in surgery and he, another resident, and a medical student came down and looked things over and told Mr. Mullen the bad news—that he needed to go to surgery soon.

"Once I get over this other thing, you mean."

"No," the resident said, "right now."

They called his wife and she said she'd be in to see him in the morning and when they told Mr. Mullen he started crying.

I stayed with him in his room while they made arrangements for the surgery. I asked him about the war. He said he'd been a navigator on bombers in Europe in World War II.

"Making daylight bombing runs into France and Germany," he said. "Like Doolittle's Raiders."

He said his life had never been the same again. "How could it be? All that time you're sure you're going to be blown to bits. We just waited to die, and a lot of guys did. Lot of nice guys."

He couldn't go on.

"I don't want to talk about it," he said. "You mind if I try to get some sleep?"

I wished him good luck and left.

A few hours later, I went up to the operating room and found them getting ready to close his wound. The chief resident told me it had ruptured and then sealed off.

"It was a time bomb," he said. "Your man was plenty lucky."

In the morning I went by to check on him and he was still groggy, but he squeezed my hand when I asked him.

I couldn't wait to tell my supervisor. I had an hour before I was scheduled to report to him, so I reviewed my notes on the other five patients and wrote down an outline of Mr. Mullen's course. I wanted to surprise him with the bit about the aneurysm. I knew from one of his lectures that depressed people often had constipation and when the colon was full of shit and lying over the aorta, it might feel like an aneurysm. I knew he'd say that, so when I would tell him what really happened and how I wasn't fooled, he'd realize I wasn't as dumb as he'd thought.

Everything went fine. The other five patients were pretty routine. I got to the part about the pulsing mass in Mr. Mullen's abdomen and he said the bit about what that was in a depressed man.

"That's what I thought at first," I said. "But—"

"Did you go over him with the resident?" he said.

I said I hadn't. I didn't tell him that the psych resident had told me not to call him unless someone was trying to kill himself.

"You need to check out all patients with your resident," he said.

I nodded. He asked if the resident had seen my other patients and I said he had not.

"Well, do it now. Mr. Mullen is likely just constipated," he said. "Now, if that's all, I have work to do."

"He had an aneurysm," I said.

He asked me what I was talking about and I told him about my exam and the X-ray, what the surgery chief resident had found when they operated, and that now Mr. Mullen was in the ICU and he was looking good and starting to wake up.

He called me negligent and said I wasn't qualified to make such a diagnosis, that what I had done was unethical. I wasn't sure about that last charge; it seemed as though he was accusing me of poaching cases for the surgeons.

He asked to see my write-up and I swallowed hard and admitted I had not written out a copy, that my workup was on the chart in the ICU and I could go write out a copy if he wanted, but that I thought this case was different since the patient was transferred to surgery before we had a chance to treat his depression. He told me I was irresponsible and that I was required to submit all work to him and that later was too late.

I liked psychiatry a lot. I thought I might be good at it if I could learn to figure people out better, but I don't think I would have fit in, and when I read my supervisor's evaluation at the end of the rotation, I was sure that was the problem. I was still naive about academic politics. I'd failed to honor the hierarchy, a system of control that sometimes overpowers common sense, if not the patients' best interest.

I appealed my evaluation to the director of student education in the psychiatry department. I told him I thought I'd pissed the supervisor off because I'd sent a patient to surgery, even though the man had a ruptured aneurysm. He said he was sure there was a lot more to it than just that and since I'd passed the rotation, he wasn't going to challenge one of his most respected colleagues.

"You can use that evaluation as a badge of honor when you apply to surgery residencies," he said.

# Cowboy Initiation

I arrived in Utah for the start of my residency in the summer of 1976. I was twenty-six years old. I quickly discovered the difference between the private world of surgery, where routine and speed meant higher incomes (and, with a few surgeons, better outcomes), and the academic world, where teaching and the greater complexity of operations were often used as an excuse for why so many cases took so long, even when there was neither complexity nor teaching involved.

I was attracted to the private world by the few surgeons who were at least as good as my dad. Even though very few general surgeons back then sought extra training after completing their residency, most of those who set up practices in larger communities ended up specializing to some degree. I discovered that one surgeon did thyroid and related glandular surgery almost exclusively, another concentrated on breast cancer, some on hernias of all kinds, others on liver and bile duct surgery, a few on blood vessel surgery, and so forth. They became very proficient by doing large numbers of similar surgeries, unlike my dad and TJ, who had to take on whatever came in the door. Specialization also provided economic advantages when a surgeon became known for expertise in his chosen niche.

I was attracted to the academic side by surgeons who refused to accept the way things always were, who looked for new ways to do things, and whose creativity and attention to detail were never stymied by that private world urgency to do more in order to make more.

The largest private hospital had some of the best surgeons. It also harbored a cast of throwbacks from another era. Many were general practitioners with no formal surgery training. They did some wacky shit. Some of it was a harmless type of crazy, like the surgeon who wouldn't let us take the dressings off the wound for three days and then did a formal sterile draping of the patient, putting on a mask and gown and gloves just to remove it and replace it with a new dressing for another three days. We were taught that the wound was pretty much sealed by the second day, and all the other surgeons wanted us to remove dressings the first or second day after surgery. I asked the chief resident about the crazy dressing obsession.

"That's probably the way they did it before Lister," he said. "Or penicillin."

Another surgeon, Dr. K, tried to prevent wound infections by laying a complicated network of tubes in the space between the muscle and the skin before closing the skin, so he could flush the wound continuously for three days with a solution containing a mix of the most powerful antibiotics available then. The rumor was he'd had a patient die from a terrible infection that destroyed the entire abdominal muscle wall and then spread everywhere. I asked him about that famous case one day and he looked at me like I was talking crazy, shook his head, and walked away.

My chief resident on that rotation was a cowboy from Texas

named Tebbetts, with thinning red hair and a closet full of Tony
Lamas. He put me in a room with a GP named Caldwell, who
had a reputation for showing up incapacitated—booze, drugs, no
one would say. We were going to remove part of a woman's thyroid, an operation I'd seen done only twice before. I told Tebbetts he couldn't leave me in there alone, that my thyroid anatomy
was a little shaky and I knew there were nerves in there that we
could injure and ruin the patient's voice. He said he'd come in at
the critical part of the operation, that he had another case to get
started. He reviewed the steps of the surgery and a bit about the
anatomy and wished me luck.

I got to the room early, scrubbed in, draped the patient, and
stood on the surgeon's side of the table as Dr. Caldwell stood
swaying at the sink, scrubbing his hands. Gowned and gloved, he
came to my side of the table and used his girth to move me out of
the way. I went to the other side and we stood there for a few
minutes not talking, and then he gestured with the sweep of an
open palm for me to begin. I used a silk thread like a garrote to
mark the skin for the incision, traced the line with a scalpel, cau-
terized a few skin bleeders, dissected the upper and lower skin
flaps off the underlying muscles, and started feeling pretty good.
Dr. Caldwell seemed content to stand there, swaying a bit, ran-
domly grunting. I got as far as putting in the retractor and exposing the thyroid gland before Tebbetts showed up and asked
Dr. Caldwell if he wanted to take a little rest. Dr. Caldwell
stepped back and the nurse brought him a stool and he sat in the
corner, his hands in his lap, his head down. He said something to
the circulating nurse, she made a phone call, and pretty soon
someone brought her a large cup and a straw, which she held to

his mouth while he took long pulls. It was white, like milk. I wondered if he had an ulcer.

When we completed all of the dissection, exposing and taking care not to injure the nerve to the larynx, we were ready to start dividing the blood vessels to the left lobe of the thyroid so we could remove it. Tebbetts asked Dr. Caldwell if he wanted to take a look before we went further. Dr. Caldwell stood, walked up to the table, bent low to look at the thyroid, reached out with his hand, grabbed it, and yanked the entire lobe out in one pull.

"Oh, Lordy," Tebbetts said.

Blood spewed from everywhere and I grabbed a sponge and held pressure while Dr. Caldwell looked at the glob of gland, turning it over and over, then handed it to the nurse and left the room.

We got things under control quickly, somehow avoiding crushing the nerve with a clamp. I waited in the recovery room until the patient woke up. Her voice was fine. She went home in two days.

I told the chief of surgery in that hospital about what Dr. Caldwell had done. He'd called me to his office to scold me for writing an order in the chart to irrigate the wound of a patient with "Dr. K's Magic Solution." He said it was unprofessional and could lead to a lawsuit. I said it was malpractice, that nothing in the literature supported putting tubes in a clean wound with little risk of infection and running all those antibiotics in there, that it carried the risk of selecting out resistant bugs and it kept patients in the hospital much longer than normal. He said that it wasn't my place to worry about that; I said it was if I gave a shit about the patients. Then I told him about my experience with Dr. Caldwell.

The chief of surgery nodded and stared at me for a moment. "And how did that turn out?" he said.

I stared back, astounded by what he meant, a kind of medical "no harm, no foul" attitude.

"Dr. Caldwell is very fortunate to have excellent young surgeons like you working with him, isn't he?"

He made me promise I wouldn't write any more orders in the chart or notes that implied anything about the wisdom or correctness of anyone's practice. I agreed. I'd simply been a smartass in the chart, and that was no place for sarcasm, criticism, or even humor.

# Geoffrey A.

My first rotation as an intern in Utah was on Red Surgery, the team led by Dr. Moody. The other faculty members were largely his protégés. My fellow intern was John Cannell. John is about six feet six or seven inches tall and I'm six five. Dr. Moody is a foot or more shorter than either of us. The chief resident on the Red service was also short. When all of us were scrubbed in together on the same case, Dr. Moody kept the operating table at a height comfortable for him and his chief resident. Both John and I had to bend over in order to hold on to any of the retractors that Dr. Moody demanded we "pull to the sky."

Jahnina was Dr. Moody's faithful scrub nurse. She ran his operating room like a Polish army sergeant, which I'm certain at one time she was. By the time I became a chief resident on Dr. Moody's service, Jahnina and I were fast friends, and we worked together to keep Dr. Moody happy whenever I was on his service. On my first day in her operating room, however, John and I were simply two more clueless interns. We were sitting ducks.

The first time I reached down to take a retractor at Dr. Moody's request, Jahnina immediately announced that I had contaminated myself and ordered me to step away from the table.

"How so?" I said.

Dr. Moody told me to do as she said and gave the retractor to John. Being even taller than I and suffering from a bit of a stiff back, John assumed a kind of squatting position in order to hold the retractor while I got a new gown and set of gloves. I returned to the table and took over John's retractor, and this time Jahnina yelled, "He's done it again." Dr. Moody asked me whether I had ever scrubbed in surgery before, whether I understood sterile procedures, and told me to re-gown and re-glove.

As the circulating nurse tied my gown, she whispered, "Don't drop your hands below your waist." I whispered back that the entire fucking operating table was below my waist.

"Figure it out," she said.

I decided that if I grabbed just the tip of the retractor handle and pulled more vertically than horizontally, I could keep my hands above my waist. That worked for a while, until Dr. Moody told me I was pulling it too high and asked me to lower it, and Jahnina went apeshit on me.

"What the hell is he doing?" Dr. Moody asked. Jahnina said I was dropping my hands below my waist.

"But the whole damned table is below my waist," I said and immediately regretted my tone.

Dr. Moody asked Jahnina if this once, we could skip that rule so that we could get the case done.

"Of course. You are the doctor," she said. "If that is your wish, who am I to say no?"

The next time I was scheduled to scrub with Dr. Moody, I got in the room early, used the platforms to create an elevated stage where Dr. Moody would stand, and asked the anesthesiologist to raise the table to a height just above my waist. I got gowned and gloved and helped Jahnina with draping the patient. I asked her what she thought about the platform.

"Dr. Moody never uses a platform," she said.

Gowned and gloved, Dr. Moody walked up to the table and without a word pushed a few platforms out of the way with his feet and told the anesthesiologist to lower the table. The circulating nurse rushed to get the platforms all removed and Jahnina never said a word about my hands being below my waist the entire operation, which solved only one of the problems. Bending over, standing with my feet splayed as far apart as I could tolerate, and trying new ways to slouch, I began to think I might need back surgery before I finished my first year as a surgery resident.

Geoffrey was one of a handful of patients the chief resident assigned to me my first day on the Red team. He was the only one of mine in the ICU. On rounds that afternoon, Dr. Moody told me that it was up to me to save Geoffrey. From what I could figure out from the chart, he was dying. He'd been treated for lymphoma with a drug that can cause scarring in the lungs and now he had to be on a ventilator all the time. The medical lung specialists, a.k.a. the pulmonologists, thought it was the chemotherapy drug, while the oncologists and Dr. Moody thought it was just a bad infection. The difference was a matter of life and death by the time I met Geoffrey. The infection might be treatable; the fibrosis of the lungs from the drug was not. The oncologists had asked Dr. Moody to cut open Geoffrey's chest and get a piece of lung so they could put it under a microscope and figure it out. The pulmonologists had already tried getting material from deep in Geoffrey's bronchioles, the small airways, and they'd poked his lung with a biopsy needle, but all that did was collapse his lung, which someone on Dr. Moody's team had fixed with a tube. The problem was Geoffrey's lung was still leaking a lot of air from that puncture, so everyone agreed he was too sick to go through the open biopsy.

The chief resident told one of the lower-level residents to teach me what to do. It seemed pretty simple, especially since I'd helped manage ventilators during one of my last medical school rotations. It was just a balancing act to keep Geoffrey's oxygen levels in a safe zone without increasing the air pressure enough to make the leak worse or to pop new holes in his lungs.

Geoffrey slept most of the time. The nurse said they kept him sedated because he'd get too agitated when awake. I came by around eight o'clock my first night on call and found Geoffrey sitting up in bed watching TV. The night nurse said he'd had a really good evening, that he woke up and didn't go nuts, so she didn't put him down.

"We've been a very good boy tonight," she said, stroking his forehead. "Haven't we, sweetie?"

Geoffrey stared at the TV and ignored her. She left and I stayed behind to check the ventilator settings. They indicated Geoffrey's lungs were better. I looked at the chest tube and couldn't see any signs of a leak anymore. I listened to his chest and his heart and all seemed OK.

"Maybe those antibiotics are starting to work now," I said. Geoffrey looked at me and raised his eyebrows. I explained the disagreement, that he was getting antibiotics because everyone agreed they couldn't hurt and might help. "You're looking a lot better tonight," I said and squeezed his forearm. It felt like a piece of wood, Geoffrey's skin drawn tight over it, seemingly without muscle or fat beneath—literally just skin on bone.

Geoffrey was my last stop for the evening. I pointed to the chair in the corner.

"Do you mind?" I said.

He shrugged. I sat down and looked around his room. Beside me on the nightstand was a packet of photos. I asked Geoffrey if I could look at them and he shrugged again. He seemed fixated by the TV. It was a war movie with Frank Sinatra.

"This your sister?" I said and held up a picture of a young woman about his age. He shook his head and mouthed that it was his wife. "These your kids, then?" He nodded. They were grade-school ages, all three. "Lot of work," I said. He smiled for the first time and shook his head.

I watched the movie with Geoffrey for a while but began to feel sleepy. I didn't want to get caught dozing in a patient's room, so I stood and said good night.

"Keep this up and we'll be getting that tube out of your throat before long," I said. He gave me a thumbs-up. That was the last time Geoffrey and I ever communicated.

Three hours later I was awakened by a call from the nurse. Geoffrey was really agitated and she'd had to sedate him. I found his oxygen levels had fallen steadily and the lung leak had increased; when I looked him over and listened to his chest I decided his lung had collapsed again. I tried stripping the tube but nothing came out: no air, no fluid. I asked for a portable X-ray and a chest tube set. He got worse while we waited. I called the chief resident and he said he was on his way.

I asked how long.

Twenty, thirty minutes tops.

I said that might be too long.

"Do what you have to do," he said.

In Cleveland, Asan had taught me to put in chest tubes, and I'd done it half a dozen times before. I got the nurse to help me and when I poked a hole into the chest with the long clamp a rush of air came out, which was a good sign that I was in the right place. With the new tube sewn in place and pulling out air that the other tube couldn't, Geoffrey began to get better. We got an X-ray and the chief showed me that my tube could have been better positioned a little more to the right. I nodded. On afternoon rounds, Geoffrey was still improving, and I mentioned that maybe he could be allowed to wake up again. "Like last night," I said. No one had known about that. They took it as good news, that the antibiotics were working.

I had the weekend off and when I saw Geoffrey Monday morning, he was dying. He had three more tubes—two in the other side and a third in the same side as mine. It was a vicious cycle in which the higher pressure needed to inflate his stiff lungs and keep his oxygen level safe led to more holes, which then leaked air into the space around the lungs, collapsing them and leading to the need for still higher pressure. The lung doctor that morning told me that the high oxygen levels we were using would make the fibrosis from the chemo drug worse.

"What you see happening here is pretty typical for that," he said.

For the next few days, no one spent much time on rounds in Geoffrey's room. The nurse called it "circling mode."

"He's circling the drain, and the doctors are circling the wagons," she said.

I frowned.

"You know. For self-protection," she said.

"From lawsuit?" I said.

"From themselves," she said. "From their fears."

I finished my note in the chart.

"You'll learn," she said. "At least you'd better."

I was on call that night and spent most of it hanging around in the ICU, twiddling with the dials on Geoffrey's ventilator, like adjusting the various dials on a sound system to get the right combination of bass and treble and volume to be loud enough but without inducing distortion. I thought there ought to be a way to make him better if I could just get the settings right. A woman in her fifties came in around ten o'clock. She said she was Geoffrey's aunt, his mother's sister. His mother had died of cancer three years before. His wife had divorced him during his early treatment for cancer and she and the kids had moved to Pocatello, Idaho, to be close to her family.

"She wanted to bring the kids down to see their dad, but Geoffrey wouldn't have it, said, 'Keep them away from me till I'm better.'"

An hour or so past midnight I went to the call room to sleep and before I left, I leaned over and asked Geoffrey to squeeze my hand. He didn't move.

Geoffrey died the next day, just before noon. I called his aunt to let her know and she gave me the name of a funeral home. The various care teams decided they wanted a lung biopsy now that it wouldn't hurt him. The chief resident had to go to the operating room and he asked if I could do it. I called Geoffrey's aunt back and got permission. The attending physician on the lung team gave me a long lecture on exactly how I should get the specimen so that I didn't contaminate it with bacteria, how he wanted me to swab the specimen for several different kinds of cultures before I dropped it in the specimen jar.

"And whatever you do, don't put it in formaldehyde," he said.

Once everyone was gone, I cut a small incision in Geoffrey's chest just below his right nipple. I popped through the layers with the clamp and spread the opening wide, then inserted my finger. I could feel the edge of something up high touching the tip of my finger and was sure it was the bottom of the lung retracted up into the chest. I worked to get a clamp on it and pull it down into the wound and then cut out a large triangle of tissue. It was much more solid than a normal lung, and a dark red color. It felt squishy, like lung that was saturated with fluid and scarred with fibrosis. I swabbed it as instructed, then put it in a sterile jar and sent it to the pathology lab.

The next day we arrived to make afternoon rounds on the surgery ward and found the lung team waiting for us, along with the infectious disease specialist on Geoffrey's case. They were pissed off. The specimen wasn't lung, they said. It was liver. A nice fat triangle of liver.

I felt a huge bolus of adrenaline rushing into my head through my neck, and my chest began to pound. Moody asked the chief how that had happened and the chief stammered a bit and said something about how he was surprised, that he'd been nearly certain it was lung. The pulmonologist who'd provided me instructions interrupted.

"He's not the one who did it. He is," he said, pointing at me, and I could already hear the tirade. An intern? You left this job to an intern? This would have cleared up months of mystery, settled the questions we all had, given us new knowledge that would have helped us the next time, save lives, solve world hunger, cure the common cold. . . .

Moody stared back at him. I wanted to point out that we'd treated him the best we could for the possible infection and he died anyway.

"Now we'll never know," the pulmonologist said. "So, that young man died for nothing after all."

They left then and we marched down the hall to the first patient's room, and as we walked the chief grilled me on exactly how I'd done the biopsy, and how on God's green earth I could have mistaken liver for lung and why didn't I call him if I was confused. Dr. Moody told us to stop talking about it and I never heard another word from anyone after that.

I wish I had. I wish Dr. Moody had given me hell for doing something so stupid. I wish he'd told me it wouldn't be the first time I'd screw up, that it happens to the best of surgeons but that I should never use that fact to let myself off the hook. Yes, shit happens, but it's still your fault. You're the one who has to be better, smarter, more careful. I wish someone else would have told me that, because that's what I told myself and it was terrifying to think I was alone.

# Jumping off Mountains in Utah

It was Tebbetts's idea to take up hang gliding. He said he'd found a guy who gave lessons out at Point of the Mountain, less than a half-hour drive south of the medical center and just across from the state prison, where a firing squad shot Gary Gilmore. And the lessons were cheap, he said. Byrd, another Texan chief resident, but with hair and penny loafers, liked the idea. "What the fuck else are we going to do?" he said. A third-year resident, Spicer, a real cowboy from Wyoming who obsessively managed an authentic handlebar mustache and made fun of Tebbetts's boots, said he was in, too. They asked me to join them. Tebbetts knew I was a pilot with an old beater of an airplane and he claimed I'd be a natural. I wasn't so sure. I'd heard that dying while hang gliding wasn't all that rare.

"Only ten percent," Tebbetts said. "No big deal."

More dangerous than open heart surgery, I thought, but I wanted to belong.

We were on rotation at the VA hospital, Tebbetts the chief on vascular surgery, Byrd on general. On a slow Friday, we made four o'clock rounds at noon and Tebbetts called up Wasatch Wings and made us an appointment with the owner, Dave Rodriguez. An hour later we were on the south side of the point, a treeless ridge jutting westward a few miles from the Wasatch

Range. We stood listening to Dave tell us all the ways we could fuck ourselves up. Behind him towered Lone Peak. Snow still lay on its flanks above the tree line and I could just make out the large ice field where only a week before a professor at the university had fallen to his death. He'd climbed up there to celebrate his fiftieth birthday. The aspen and scrub oak along its lower slopes were just starting to change.

Dave had two kites set up. He said one was "real basic" and the other a bit higher performance. Other than color I couldn't tell the difference. We took turns carrying a kite a short way up the hill, clipping in, and running down the hill until the kite pulled us off the ground and we flew maybe ten feet and crashed. After half an hour, we weren't crashing, so Dave had us walk another hundred feet up the hill and now the flights lasted maybe fifteen seconds, which felt like forever at first. By the end of our third or fourth lesson, we were all taking off from the top of the hill, three hundred feet high, and doing three-minute flights. Dave suggested we try soaring a bit by turning perpendicular to the ridge and staying in the lift coming from the southerly breeze flowing up the slope. In no time we realized we could stay up as long as there was enough wind coming out of the south. We could even land back on top so the next guy didn't have to carry the kite back up the hill.

During the week, we often had the hill to ourselves. Weekends brought out a diverse crowd of flyers, from folks with advanced flying wings to guys with original Rogallo kites—the plain triangular wings that had a reputation for blowdown. That's what Dave called the problem when the pilot lowered the nose a little too much and the wind hit the top of the sail. He used his hands to show the wind pushing the kite into a steep nosedive and exploding on the ground. He showed us how the more modern kites we were using were designed to prevent blowdown.

When the south winds were right, the weekend skies above the hill could see a dozen kites working the lift. On a clear and sunny Sunday afternoon late in September, three of us were practicing our soaring on the south hill, taking care to follow the traditional elliptical course of going west along the ridge close to the hill and coming back just far enough out to avoid hitting anyone coming west. Sometimes we'd get a bit too far out from the hill, lose the lift, and have to land at the bottom and carry the kite all the way back to the top.

That happened to me on my second flight and I could hear Tebbetts yelling at me from the top. He was pissed because it meant a long delay before his next flight. As I started up the hill with the kite on my shoulders, a guy pulled up in a rusted-out Datsun pickup and set up his kite. It was the first old-style Rogallo I'd ever seen up close. It was small compared to the gliders we used. It looked homemade. He started up the hill and made it to the top nearly ten minutes ahead of me. I saw him sitting on the ground, arms resting on his knees, staring out to the south. He had a ponytail hanging to the middle of his back and a long beard and was wearing a Moody Blues T-shirt, cutoff jeans, and sandals.

Byrd and I were watching Tebbetts soar back and forth in the train of five other hang gliders when Mr. Rogallo decided to take a ride. He lifted the kite to his shoulders and ran. He lifted off, flew straight out over the valley, the angle of flight getting steeper and steeper until the wing stalled, the nose dropped sharply, the fabric blew down, and he crashed into the ground from about a hundred feet up. Byrd and I looked at each other and then sprinted down the hill.

The pilot was in the midst of a grand mal seizure when we arrived, and then he went completely still. He didn't have a helmet; his nose was smashed and the side of his face was abraded and bleeding from where it had hit the dirt. Both collarbones were snapped, the shattered end of one sticking out six inches through his skin. Someone said they had a radio and called for help. Byrd and I took turns holding his head straight and his jaw forward to keep his airway open. The pilot was still unconscious, but breathing, when the ambulance arrived.

A guy who'd talked to him up on the hill said the pilot was from California. The kite had been his brother's and his brother had recently died of cancer.

"He said he'd only flown it once before," the guy said. "But he was pretty vague about that."

A few days later we found out he'd survived and was waking up slowly. I asked if his neck was broken, but no one knew.

We bought our own kites that fall, three in all. Byrd and Spicer shared one. I ordered mine with forest-green fabric and a yellow stripe on the right wing. I couldn't afford it, and my wife reminded me of that, but I did it anyway. It helped me feel alive. I was among guys who shared the thrill, who understood why we had to do it.

We learned to fly off the north ridge, which at fifteen hundred feet was five times higher than the south hill. Landing on top was much more difficult, and for a while we just landed at the bottom. But that meant someone had to drive the truck back to the bottom, a twenty- to thirty-minute trip. Tebbetts decided that was a good job for the medical students on rotation with us at the VA.

The first student we took thought it was very exciting. He said it was much better than sitting around at the VA waiting for a resident to tell him to go dig shit out of some guy's rectum. We must have pissed off one of the students, though—enough to turn us in. Dr. Moody called Tebbetts and Byrd to his office. The student claimed Tebbetts had threatened him with a bad grade if he didn't go with us. Tebbetts said he was only kidding and thought the student understood that. I wasn't sure Tebbetts was kidding, and I worried it might come back to bite us. Soon after that, we learned to land on top of the north ridge, so one of us could drive the truck down and pick the others up in the valley.

Winter came and the days remained warm and sunny, so we were soon traveling to other locales to jump off more interesting places. One weekend in January, we soared for hours with bald eagles among the rocky spires of a five-mile-long ridge on the Wyoming border near Randolph, Utah. Several times we jumped off a nearly ten-thousand-foot mountain called Francis Peak, north of Salt Lake. On one trip, I flew too far along the Wasatch mountains, lost the ridge lift, and ended up in a vacant field six miles south of our designated landing spot. My friends found me two hours later. They were more pissed off than happy to see me.

We had bigger plans that we couldn't pull off. One was a flight off Mount Timpanogos in the winter. Tebbetts thought we could launch from the saddle that sits at about 11,250 feet, or 500 feet below the summit. He said it was only about an eight-, maybe

nine-mile hike and not more than four, maybe five thousand vertical feet of hiking. And the snow? I asked. He said we'd use
snowshoes—or skis; if we got up there and couldn't launch, we'd
have a hell of a great ski trip back down. We'd tow the kites on
sledges, he explained, not our backs. "Oh," I said, "that makes
sense." I said it was a great idea. I knew we'd never get around
to it.

Tebbetts and Byrd finished their residency that summer and
Spicer no longer seemed interested in flying. In the fall my wife
and I drove down to southern Utah. We took my hang glider
because I had this idea of jumping off at Dead Horse Point. We
camped in Moab and explored Arches National Park while waiting for a clear day with no wind before driving out to the point
early one morning.

While my wife went off to explore the spectacular views of
the Colorado River, I stood on a launching ramp that a national
hang gliding organization had made to help with launching off a
cliff during its annual contest. I'd never launched off a cliff.
Dave had told us something about it, but that ramp was about
five steps long. I wasn't sure where other gliders usually landed.
I was hoping it would be obvious. I could see the road fifteen
hundred feet below on the white rim. The Colorado River was
another five hundred feet below that and the cliff looked very
close to the road. I also worried about all the rock outcroppings
I'd have to soar over, or between, depending on how much lift I'd
get from wind coming up the cliff. I picked up a broken branch
of sage and tossed it over the edge. It flittered down a few feet
and lodged in a crevice. Not much wind at all, I thought. That
just made the idea of a cliff launch scarier. From what I recalled,

no wind meant a longer dive before I'd have enough lift to start flying.

"Thinking about it?"

I nearly jumped off the ramp. A park ranger stood behind me, smiling.

"Sorry," he said. "Didn't mean to spook you."

I said I wasn't sure, that the wind seemed awfully calm.

"Is that road there where they usually land?" I asked.

"That's one place," he said. "Over there's another." He pointed off to the north and I thought I could see a flat spot on the other side of a big rock pillar.

"I think it depends on the wind," he said. "Something about avoiding rotors off those rock columns and those outcroppings over to the south. You know, like little whirlwinds you can't see."

I could only stare.

"You do this a lot?" he said.

I said I was pretty experienced, mentioning a few places I'd been. "Never done a cliff launch like this, though."

He didn't say much and my wife came up, handed me a candy bar, and introduced herself to the ranger.

"What do you think?" she asked me.

I said I wasn't sure. I asked the ranger how long a drive it was from where we were to that landing spot down on the road. He said it might only take an hour and a half, but it could take up to three if you got behind a slow truck, or if you weren't used to that kind of driving. He said the Shafer Canyon road was pretty hairy.

"Dirt road with lots of loose switchbacks and no guardrails," he said.

I thanked him for his time, walked back to the truck, and sat in the cab, trying to decide what to do. I was scared, more scared than I thought I'd be. I hadn't flown for a couple of months and the cliff was so freaking huge and I didn't think I knew what I was doing. On the other hand, it could be spectacular. The day was still young and if all went well, we'd be back in Moab toasting a great day long before dark. And I could say I'd done it. Not many people knew where Francis Peak or Camelback Mountain were, but Dead Horse Point was pretty famous.

I had a bad feeling and it kept ramping up my fear level. I put my hands on the steering wheel to try to stop the shaking I felt. My wife had disappeared onto another trail. She had to leave me alone.

I made up my mind and grabbed my camera, and by the time I found her I was nearly ecstatic.

"Well?" she said.

"Fuck it," I said.

"Good. I wasn't sure I could do that Shafer switchback road thing he was talking about."

By the summer of 1978, flying the north ridge out at Point of the Mountain had become a scary proposition. One Saturday afternoon, I counted twenty-nine ships in the air with me, nearly three times the number we used to get, and they were all vying for the best lift. And no one was following the tradition of flying an elliptical pattern. Guys were dive-bombing me to try to get into some lift I was already using. The landing area on top was too crowded for me to feel safe, so I sent my wife down to the bottom to wait for me. At ground level the temperature was in the nineties, so I climbed and climbed until I was well above the other kites and the air was freezing cold. I couldn't afford an altimeter, but I figured I was up close to fourteen thousand feet, maybe more. My head began to hurt and I felt short of breath.

With Tebbetts and Byrd gone and Spicer reluctant, the camaraderie was gone, replaced by crowds of crazy pilots from who knew where. I was no longer part of a group that accepted me for who I was, where I could jump off a mountain under a Dacron wing and feel like I was king of my own domain. Taking time off was also becoming harder as the rotations got busier, and spending what free time I had hang gliding in solitude seemed utterly selfish. Even so, I couldn't escape the sense of power that came with every flight, bald eagles alongside or not.

On a Friday afternoon in early October, I talked a friend named Steve into driving me up to Francis Peak. The forecast winds were about right, possibly getting stronger—maybe too strong. I'd decide when we got to the top.

We arrived at the launch site by four o'clock and I got set up and ready to clip in quickly. The winds were still building, and a line of clouds appeared suddenly out of the west. I was scared. I worried about a front coming, bringing high winds that would blow me and my kite backward over the summit and into the rotor. I thought about my failure at Dead Horse Point, about the regrets I'd harbored since about not jumping. Besides, I'd jumped off this peak before and here I was again, all ready to go. We'd come all this way and I was convinced that once I got away from the slope I'd be fine.

I explained to Steve that with the winds the way they were, I needed him to get out in front of me, slightly downslope, and hold on to the front of my ship. I told him that when I said go, he should let go and drop to his hands and knees, off to the side.

I took a few deep breaths and tried to talk myself out of it, but then I blurted out "*Go!*" and started running down the slope. Steve was slow to get out of the way and just as I became airborne, my left wing tip caught his shoulder. The kite rotated toward the left and I couldn't get it to stop banking that way; I was suddenly headed straight for the rocks.

I'd never crashed before. Tebbetts had crashed once and did a number on his ear, but that was because he'd set the kite up wrong. I'd watched it happen from way overhead, and when he didn't move for a while and I thought he was dead or dying, I screamed for someone down there to help him. My crash wasn't so dramatic. I got bruised and scraped a bit, but that's all. My kite wasn't bent up, but two of the aluminum tubes were ruined.

Steve helped me haul it back up the slope and pack it up. I didn't know that would be the last time I ever went hang gliding.

I knew about twenty guys who flew regularly, and one of them died a few years after I quit. I never found out how. Dave, our instructor, got caught in a rotor behind a mountain and it made his kite do two complete loops before he landed on his butt in the rocks and broke his back, but I think he was mostly just sore for a long time.

One late night a year or two later, I found myself talking to a surgeon from Spain who had taken up paragliding and I told him I'd done a lot of hang gliding. I told him about the day I jumped off Dead Horse Point and glided for nearly an hour before landing on a road next to a cliff five hundred feet above the Colorado River. I think I started believing in that lie after that, and I retold it whenever the subject came up in casual conversation. I might have written that false version of the story here had I not realized it was a lie I couldn't tell anymore, not only because I knew it was dangerous and unethical but also because the truth is far more interesting. I didn't lie to others to brag about myself, nor out of fear of being found a coward. I lied to protect myself from a far greater reality: I backed away from the cliff because on that day, I recognized I couldn't control the outcome. In the long term, however, I couldn't accept the notion that there was anything I couldn't control. I rewrote my memory of the story to restore my self-confidence, to avoid facing how little control I ever have over anything. In my lie, I jumped not because I didn't want to seem cowardly, but because not jumping meant I didn't have complete control.

I never told my dad that I hang glided in Utah. I didn't want to face his criticism, his calling me foolish, his rants about the dangers of such a reckless thing. By the time I regretted not telling him, it was too late—he wouldn't have understood what I was talking about, nor cared. I still have my kite, along with a couple of different harnesses and a helmet. They're in a storage unit. Someone might want them.

# Fired

During the fall of 1977, my second year in Utah, I rotated to a hospital forty-five miles north on I-15, in Ogden. I was there with another resident from my year and Spicer, then in his fourth year. The surgeons we worked with were very busy and we worked hard to cover all of their patients. The resident on weekend call arrived Friday morning and didn't go home until Monday evening. The cafeteria food was good and sometimes we had some interesting weekend cases. The call room had a color TV, a big soft couch, a full-size bed, and a refrigerator.

On a Monday morning after one of my weekends on duty, I was writing orders on my patients before going to the operating room when the ward clerk leaned in and in a quiet voice said, "I'm so sorry to hear you won't be rotating back with us next time." I laughed and she didn't, so I asked her what she was talking about. She looked embarrassed and said I needed to talk to Dr. Spicer. I found Spicer at the scrub sink and asked him what the clerk was talking about. He said he'd talk to me between cases and reminded me I was supposed to help Grua with a colon in room five. "You don't want to keep him waiting," he said.

I'd been fired. Dr. Moody had decided to reduce the number of residents in the program and I'd fallen into a group of six guys who would either be let go immediately or be allowed to stay the

rest of that year or until they got another residency spot somewhere else. I called Dr. Moody's office and his secretary said he
was out of town for the week. A close friend of mine was at the
meeting where it had happened. He said the voting had been
close, that I'd gotten some bad press because I'd hung out with
Tebbetts and Byrd too much during that hang gliding nonsense,
that some of the attending surgeons felt I was a little too confident, and that one said I didn't know anything about the thyroid
gland and that I could be a bit of a troublemaker.

"Other than that," he said, "you did pretty well."

My wife and I were all set to take a vacation the week that Dr.
Moody was due back. I thought about canceling the trip, but
some of my fellow victims said he had told them the decisions
were final. I didn't see the point of sticking around. We drove to
Portland, Oregon, to go hiking with Paul, a college roommate,
and his wife. We fell in love all over again with the Pacific Northwest and I got this great idea that we'd move to Tenino, Washington, where Sorrell Aviation had built this wonderful biplane that
I'd long dreamed of flying. I'd work weekends as an emergency
room doctor in Olympia and the rest of the week I'd work for the
Sorrells doing whatever they asked in exchange for their help
and materials to build my own Hipcrbipe, as they called it.

At the end of our time with Paul, we drove to Tenino and one
of the younger Sorrells said it might work out, but that the old
man was away for a week. In Olympia I picked up applications
for two hospitals. One administrator said they were desperate for
weekend coverage. We left for home excited about starting a new
life in the Northwest.

On the Monday after we got back, I found out it was all off.

Dr. Moody had changed his mind and we all got our jobs back. I met with him and he told me what I must do to get better evaluations. Mostly it had to do with being less of a smart-ass and more diligent about reading the surgical literature so I wouldn't seem so ignorant.

My wife and I talked about my quitting anyway. We'd been so happy with our plan. But I also felt tremendous relief, like I'd been given a second chance.

Late in my second year, I decided I wanted to spend my third year of residency in Dr. Moody's research laboratory and got my approvals in order three months in advance. Then the faculty had their annual resident evaluation meeting and once again, my performance was judged marginal. They canceled my year in the lab and told me that I'd have to do a lot better if I expected to finish the program. My new schedule didn't include any time on Dr. Moody's service, so I traded with one of the other residents for three months with the boss. I found time after-hours and on weekends to work in the lab anyway and by the end of my three months, Dr. Moody gave me another month in the lab, but free of any clinical duties. That spring, I got a paper accepted to the American College of Surgeons Forum and a free trip to Chicago to present my data, only to have my ass handed to me by a famous surgeon from Galveston who knew a lot more about what my work meant. Still, it was my first surgical publication, and by then I'd realized I wasn't going to work for my dad when I finished my training. I wanted to join the faculty of a university and teach medical students, train surgery residents.

# Janie and the Giant Abscess

Gary Maxwell was chief of transplantation at Utah. By the time I rotated onto his service during my third year, he was also a defrocked Mormon bishop, divorced and living apart from his large family, and as far as I could tell, a happy man. He also became one of the most influential men in my development as a surgeon.

Dr. Maxwell taught me how to take care of kidney transplant patients, and in doing so, made me feel valuable, respected, and part of a team. My opinions weren't always that useful, but he never treated me as though I were a lower form of life. Far from it—he took time to explain what he was doing, and was mostly patient when I asked for explanations. Maybe most important, he taught me how to be suspicious of anything amiss in those very fragile transplant patients.

In the late seventies, we had only two approved drugs to keep the immune systems of transplant patients from recognizing the newly implanted kidneys as foreign objects and destroying them, a process called rejection. These two drugs were like shotguns. We aimed them at the transplant patient's immune system and pulled the trigger and generally fucked them up good. They had trouble fighting off infections or healing their incisions or maintaining their strength. The idea was that by hitting them hard right when you put in the new kidney, you could convince the immune system to give up and accept the new organ as a harmless boarder. It was always an uneasy truce, though, and a week or two later the immune system would have second thoughts and you'd have to punish it with another shotgun blast.

All of that would change by the time I moved to Pittsburgh, when, for the first time, we'd have a new drug that was much more like a sniper rifle, a weapon with which we could target only certain parts of the immune attack, leaving most of its useful functions intact.

In 1978, Dr. Maxwell taught me to trust nothing about the transplant patient's body signals. Wait for a fever or a high white blood cell count to diagnose an infection and you'd often be way too late. Rationalize that the patient couldn't possibly have a great big old abscess hiding in her belly because she had no pain and you might only find it at her autopsy.

Janie was a twenty-three-year-old with kidney failure from advanced diabetes who got a kidney transplant a few months before I rotated onto Dr. Maxwell's transplant team. She was back in the hospital because she had night sweats. Her new kidney was working fine, and most of the time she felt good, though she admitted she'd been getting steadily weaker for the past two weeks. Both ultrasound and CAT scan technologies back then were pretty crude, and we had several studies using each that didn't show anything that looked like an infection inside her abdomen or in the cavity near the groin, where they'd placed the new kidney. Her blood and urine and throat cultures didn't grow any abnormal bugs, and she didn't have a cough or a stuffy nose or diarrhea.

Janie wanted to go home. She said the night sweats weren't that bad and maybe now that the weather was better she could get out and walk more and get her strength back. I thought that seemed like a good idea, but Maxwell wanted her to stay a little longer. Just in case, he said. In the hallway he told the nurse and me that something wasn't right. He stared at the floor for a moment, then moved on to the next patient.

On my morning rounds a few days later I found a small abscess in Janie's groin. About the size of a walnut, it was red and soft and Janie said it barely hurt at all. I asked her if I could numb up the skin and lance it. Like a boil, I said. I thought this could be what was causing all the trouble and it had finally shown its face.

I got all the stuff I needed together as quickly as I could. I wanted to drain it and have it all taken care of by the time Dr. Maxwell and the rest of the team came in. I'd drained dozens of abscesses before, so it was a minor procedure for me, but Janie was far calmer than I was. I felt like a forty-niner who'd just found a nugget in his pan.

I sterilized the skin and laid out the drapes as Janie's nurse handed me what I needed. I numbed her skin and with a small scalpel blade made an inch-long incision over the lump. I had a syringe in hopes I could collect some of the pus to send to the lab for culture so we'd know what bugs were in there, but when I popped through the skin into the abscess the fluid shot out like water from a garden hose, soaking my shirt and tie. I immediately applied pressure with a sponge and stood there terrified about what I might have done. Had I cut into some sort of weird hernia and opened up the intestines? Or the bladder? It looked like pus, only thinner. Maybe I didn't know something critical about where they'd placed the transplant kidney, a detail I should have known before cutting holes into things.

I asked the nurse to get me a basin. When I took off the pressure to let it drain again, I directed the flow down into the basin. It was pus, lots of pus. It kept coming and coming. The nurse got a second basin when the first was nearly full and before it stopped, we'd filled both nearly to the top. I guessed it was a good two liters, maybe more. When the flow ebbed to a dribble, I packed the wound with some gauze and went to call Maxwell.

We took Janie to the operating room that morning and opened the incision wide enough to gently wiggle my hand in and feel around. The abscess was so huge that even with my arm all the way in up to my elbow, I couldn't feel the upper limit of the cavity. I could feel the transplanted kidney; the abscess had formed around it, then over time expanded its volume all the way up to her spleen. We got another gallon of pus out and rinsed out the cavity with liters and liters of fluid, then put in a couple of two-inch-wide rubber drains that looked like long, flat, hollow strips of pasta and sent her to the ICU.

Someone with a healthy immune system would have been sick as a dog by the time that thing had grown to the size of a baseball, if not before. A healthy person might have gone into septic shock and taken weeks to get better after the kind of surgery we did. But not Janie. Janie had been treated for at least three separate bouts of rejection with massive doses of steroids—not the kind weight lifters use to build muscle; the kind that weakens muscles and talks white blood cells into ignoring bacteria.

Janie got about as sick as a princess after a pedicure. She

never had a fever after surgery or any signs of an infection, and a month later when she came back to the clinic, her kidney was still working fine and she said she was up to two miles on her morning walks.

I worried Dr. Maxwell would be angry I hadn't called him before I drained the abscess. I worked out in my head that he'd tell me I was incredibly irresponsible, that I had no business cutting people open on the wards. I decided I'd just take it and apologize and hope to feel at least some solace knowing that I'd found out what was wrong with Janie. On his way out of the operating room that morning, though, he stopped and looked back at me.

"I wish I'd seen the look on your face when that pus started coming out," he said. "Oh, and the shirt and tie are on you. Consider it your consolation prize."

# Burned Yolanda

Mike Duff was the senior resident on the burn service when I was an intern. He was a Rhodes scholar and studied physiology while at Oxford. He grew up in the Ozarks and had an accent that was thicker than the one we affected in southern Ohio. He looked vaguely like Alfred E. Neuman but with a small mouth and curly red hair, and he was the smartest person I'd ever met. Mike rode a Norton 750 to the hospital and walked in the front door in full leathers carrying his helmet. Sometimes he took me to lunch at a strip club. One of the dancers was a striking Vietnamese woman with an obvious appendectomy scar. Mike saw me looking at it. "Wound infection," he said. "Appendix ruptured by the time she came in. One tough broad, that one."

On my first day on the burn unit with Mike I met Yolanda. She was in a coma and on a ventilator. She had deep burns over nearly half her body. She also had severe lung damage. The chart said that she weighed more than 300 pounds, but she had been closer to 150 pounds when she arrived. I looked back in the chart and saw that during the first few days of her resuscitation, they had given her more than 70 liters (or about 18 gallons) of fluid by vein. About half of her 300 pounds was fluid that had seeped into her tissues and lungs.

When Mike and I made morning rounds together, we thought

her biggest problem at the moment was the fluid in her lungs. Her kidneys seemed to be working well, so we thought we could help her breathing if we got some of the excess fluid off. We'd have to be gentle, reducing the amount of fluid in her circulating blood in order to draw fluid in from the lungs and tissues without getting her too dry and damaging the kidneys.

Mike told me the head of the burn unit and his partner were both out of town at a national burn meeting. Our backup was a surgeon in private practice who used to work in the burn unit. After we'd seen all the patients, Mike called the backup surgeon to let him know our plans, including our ideas for getting fluid out of Yolanda.

We made evening rounds with the backup surgeon and I reported that our plan for Yolanda wasn't working. Dr. Backup said there really wasn't much we could do but give her more time.

On call in the hospital that night, I went to the library and looked for articles that might give us other ideas for Yolanda. I found something I thought might work. The authors of one paper wrote that it was like recharging the cells so that they stopped being so leaky. It was a new idea and all I could find were anecdotal reports of success in patients like Yolanda. I told Mike about it the next morning. He read the papers and thought it was a reasonable idea (later research would prove the technique was ineffective in changing the outcome for patients like Yolanda). "Probably won't work, but it's low risk and we don't have any other ideas worth a hill of beans."

We got started about noon, recording in a tablet the weight and the urine output before we started, then doing the slow injection, carefully monitoring everything to make sure we weren't doing any harm. By evening, we saw her urine increase and so we repeated the injection. By the second day, we'd gotten about forty pounds of fluid off. Her blood pressure was good, and though her blood tests suggested we were making her circulating blood volume a little dry, we didn't think that was worrisome. I was on call again that night and went to see Yolanda around midnight. Her blood tests were the same and she was still making a lot more urine than before. The director of the unit was due back the next day and I was excited to show him what we'd accomplished.

I overslept the next morning and didn't get to the burn unit until nearly seven o'clock. The director was already there, sitting at a desk, furiously writing in the chart. The nurse said he'd come in at four a.m.

"He heard about your little experiment," she said. "He's not too happy."

I wanted to wait until Mike arrived, but I knew the director had already seen me, so I took a deep breath and walked over.

"Welcome back," I said.

He finished writing a paragraph or so, laid down his pen, and turned to face me. He said what I had done was as bad as anything the Nazis had ever tried. He told me this would be the last straw for me, that he was going to report me to Moody and I'd be lucky if I wasn't brought up on ethics charges, that if he had his way, I'd never be allowed to practice surgery.

Moody called a meeting later that day. Mike and I collected the papers I'd found in hopes we could prove our idea hadn't been either crazy or unethical. Mike said to let him do the talking, take the heat for it, claiming he was less vulnerable than I.

The director was already talking to Dr. Moody when we arrived and he seemed at least as livid as he had that morning. Dr. Moody asked Mike what the hell we'd done.

"Well, Dr. Moody, it was just a little ex-spear-ee-mint," he said.

That's it, I thought. We're fucked.

The director nearly choked on his coffee. He said someone needed to protect his patients from negligence like that. Dr. Moody turned to me and asked what I had to say. I handed him the stack of papers from the literature and told him I thought what we'd done wasn't crazy, but a legitimate form of treatment for patients like ours.

"And it was working," I said.

I handed him the tablet and pointed to the urine output and weight and the blood tests reflecting kidney function.

"She's lost about sixty pounds," I said. "As of midnight."

The director argued that the blood tests showed she was dangerously dehydrated and that he had to give her nearly thirty liters of saline just to get her back into proper balance.

The meeting didn't last very long. Dr. Moody didn't say anything to us in front of the director but later called us back alone and chewed us out for not keeping an attending surgeon involved.

"You can't go off on your own and try something like that without getting the full approval of your attending. That will get you fired."

I wanted to tell him we had checked it out, with the backup

surgeon, who had told us he thought it couldn't hurt, but Mike and I had decided to leave him out of it. Mike tried to claim it was all his fault, that the plan had been his idea.

"Stop the bullshit, Mike," Moody said.

We still had the better part of two months to go on the burn service and I never felt like the director forgave me. He let me do the usual stuff in the operating room that someone at my level could do, but I felt like I was never in a position to ask questions about any of his plans for the care of the patients in the unit. By the morning after our meeting with Moody, Yolanda had gained all sixty pounds back plus another ten or so. She was starting to wake up but was still on a ventilator and still swollen like a poisoned toad when my rotation ended. I later learned she eventually died of infection. Her chances of recovery were never good, no matter what we would have done.

# But Why Bother?

After finishing his residency in 1979, Mike Duff moved back home to Houston, Missouri, a town of about two thousand souls. Houston is the county seat of Texas County, population twenty-five thousand, and surrounded by the Mark Twain National Forest.

Mike sent me a letter the next spring describing the intensive care unit he had set up. I was amazed. He'd created a state-of-the-art ICU in a part of the world more backwoods than my hometown in Ohio. I was still torn between going into transplant surgery and returning to Ohio to work with my dad and TJ. I decided that on our way to visit Dad and family in Ohio that summer we'd take a detour south to visit Mike and his hospital. If I was going to go back to Washington Court House, I wanted to bring something new to the area. I thought Mike could give me some ideas about how to make it work.

Our visit to Missouri featured eluding the sheriff who chased us through the Mark Twain National Forest for speeding; eating pulled pork and dancing the polka at a celebration of the National Trail Ride; and helping Mike's wife put up a fence to keep the copperheads out of their daughter's play area in the backyard. I toured Mike's ICU and heard from grateful patients who were convinced Mike was the Second Coming. I was dumbfounded by how much he'd accomplished in such a short period of time. I was excited to talk to my dad about my ideas for his hospital.

After dinner in Ohio a day or so later, Dad asked me to take a look at a patient. One of the general practitioners had asked him to see the man. Dad thought the guy had heart failure but some things weren't right with that diagnosis.

I tried to rouse the man by talking to him, then by pinching his skin.

"He's not very responsive," the nurse said.

I asked if the patient drank.

"Definitely," the nurse said. "He also stopped making urine this morning."

I pulled back the sheets. The man's belly was full of fluid and I realized then he had cirrhosis and that his liver was probably failing. I figured we really couldn't do much for him.

I sat him up and watched as the blood level in his jugular vein fell.

I said maybe some more fluid would help.

"Hell, he's already full of fluid," Dad said. "Look at all that swelling. Everything we give him just leaks out. I've been giving him diuretics to try to get fluid off."

I told him the guy had liver failure and probably kidney failure because of all the diuretics. I said there wasn't really much to do.

"I was afraid of that," Dad said. He stood looking at the patient for a moment. "Well, thanks anyway."

On another evening, I helped him take care of a guy who had heart failure. I showed him how to put a catheter into the large vein just above the heart in order to make sure he gave him just the right amount of fluid but not so much to push the guy into further heart failure. It was the first time anyone had done that in his hospital.

Later that same evening, I found Dad in the den in his vinyl lounge chair. He was watching *The Rockford Files*, or rather he was sleeping while the TV blared a chase scene with James Garner behind the wheel of his Pontiac Firebird. I loved that show.

Dad woke up when I sat down on the couch.

"You're missing a good one," I said.

"Just dozing a bit."

"Right," I said. "So who's he chasing?"

"What?"

We watched for a while and I noticed Dad starting to nod off again.

"You been working a lot?" I said.

He looked blankly at me, his lids drooping. "Oh shit. Every night. TJ's out of town for two weeks."

"Horse show?"

"Hmmm."

The show was starting to wind down. I thought he might skip the news and go to bed early.

"Do you remember Mike Duff?" I said.

"Mike Duff . . . ?"

"We had dinner with Mike and another resident last year when you came out to go skiing. He's the Rhodes scholar, the baseball pitcher from the Ozarks. Talked like a hick, worse than me."

"Worse than *I*."

"Exactly," I said.

He looked back at the TV.

"He finished last year and went back home and set up a surgical practice and an ICU. In a town less than a sixth the size of Washington Court House."

"How 'bout that."

I told him we'd stopped by in Missouri on the way home and saw the setup and how important it is for the people in that area.

"Hell, one time he sent a patient to Springfield and they called him to ask what kind of catheter the patient had."

I waited for him to ask but he was entranced by James Garner. Earlier in his career my dad's favorite show had been *Maverick*.

"It was a Swan-Ganz catheter. He sent them a trauma patient with a bunch of broken bones and a pneumothorax that he'd stabilized in Houston in his new ICU."

"I thought you said Missouri."

"Houston, Missouri."

"There's a Houston in Missouri?"

"Yes. In Texas County."

"How 'bout that."

The credits started to roll.

"I was thinking that when I come back in a couple of years we could do something similar, only bigger. We could accept patients from Greenfield and Hillsboro and even Wilmington. Set up an ICU to take care of patients like that one we saw this evening, maybe stabilize more serious ones before shipping them to Columbus. What do you think? Would the hospital go for that?"

"Oh, I suppose we could, but why?"

"Why?"

"I've been here for almost thirty years now without all that stuff."

"Well, I guess because—"

"I figure they're that sick, we can get them in an ambulance and up to Columbus quicker than you can put in one of those swan-things."

"But I'm sure some of them probably die on the way and—"

"Hell, Ohio State bought one of those helicopters. They were down here last week during the fair giving people rides. They say from the moment we call them they can be down here in less than twenty minutes."

I didn't know what to say. Maybe he was right. He had a big university hospital less than forty miles away. Why complicate things?

"Besides," he said, "by the time you get here, TJ and I will have a hell of a time teaching you the right way to do things. You'll have a lot to unlearn after those professors get done with you."

He got up then and went to the bathroom. I went out to the kitchen and got a bowl of ice cream. My wife was drawing at the kitchen table.

"How's it going?" she said.

"Great," I said. "He's just about there."

When I got back to the den, he was asleep again, snoring. I turned off the TV. My dad was six months from his sixtieth birthday and he was so tired from working at night while his partner was out of town that he couldn't stay awake long enough to have a conversation, or watch his favorite TV show. His was a lonely life, I thought. With his partner out of town, he was the only surgeon in three or four counties. He did the daily list of operations in the morning, sometimes into the afternoon, saw patients in his clinic four afternoons a week, then made rounds in the hospital each evening. He sent the patients with injuries or problems that he and the hospital couldn't handle to Columbus, but for everything else, he had no backup. I didn't see myself getting stuck there. I didn't want to be so alone. I wanted to be part of a larger team.

At first, I thought his opposition to my new ideas a resistance to change. He was too comfortable doing things the same way for too many years, I thought. Yet over the next week or so, what he said began to make sense. I knew that consolidation of specialized medical services in major urban medical centers was already happening, that it was the right way to deliver the kind of care that was rarely needed in smaller communities. The patient he had asked me to see didn't need an intensive care unit that might treat a handful of patients every few months; he needed either a liver transplant or, more likely, comfort measures to allow him to die with dignity.

I'd been captivated by the idea that I could return to my hometown with new skills and be the kind of hero there that Mike Duff was in Texas County, and a different kind of hero than my dad and TJ. I'd have my own thing. Only years later did I realize that my ambitions were driven more by my expanding ego than by the needs of that community. I had no reason to blame Dad for my not coming home.

# From Best to Worst
# and in Between

Seventeen years to the day after my father married the drunk who would become my stepmother and later teach me to be a scrub tech, I got to do my first liver transplant, from start to finish. Starzl picked what he called a perfect case, a younger patient with no previous surgery and a liver that was perfectly normal except for a missing enzyme, a flaw that threatened his life. It was like operating on a normal liver, with none of those boggy veins ready to burst and flood the field with blood. Dr. Starzl let me start the operation with Shun while he flew out to get the new liver.

When he came back, he and a visiting surgeon fixed up the new liver in the basin full of ice water. Once done, he came over to see if we were ready. Shun and I removed sponges and lifted and turned the liver this way and that, until Starzl said it was good and went back to the table. "Let me know when you're ready for this," he said.

Shun watched me put on the clamps and made sure I cut the vessels plenty long enough for the new liver. We handed the old liver off to the scrub tech and spent a few minutes sewing up some minor bleeders and making sure we had good cuffs on all the vessels for attaching the ones on the new liver.

"OK," I said. "Let's have that liver."

"Yessir," Starzl said. "Coming right up."

I heard a loud splat on the floor behind me then, a sickening sound I'd never heard before but recognized immediately.

"Oh no!" Dr. Starzl said. "What have I done?"

I turned and there on the floor in front of him was a liver, the new liver. I felt my knees go weak. I reflexively asked for some Betadine, the reddish brown iodine stuff we use to sterilize our hands and the patients' skin, thinking we could salvage this. We just had to dust it off. Like a ten-second rule for donor organs.

Starzl and his visiting surgeon looked at each other and laughed and I thought them insane until the visitor pulled back a towel from the basin and lifted the real liver up to show me.

"Fucking hysterical," I said.

Starzl's affect changed immediately.

"Come on now," he told the visitor. "We're running out of time." He grabbed the liver, carried it up to the operating table, and forced his way between me and an assistant on my left. I asked for the first suture and suddenly I was sewing in the new liver.

I found myself struggling to see, and having a difficult time getting into the place where I wanted to be because Dr. Starzl was in that place, watching my every move, verbally guiding every stitch I placed.

"More to the left," he said. "No, less. There!"

He kept after me to be quicker.

"Come on, now," he said. "This is the place where you can really make up time."

Every step became a chance for me to make up time. The

room felt hotter and hotter and by the time I did the last connec-
tion and told the anesthesiologist we were going to release the
clamps, I was soaking wet and shivering in spite of the heat. I let
the clamps loose, restored the blood flow, and watched the new
liver turn pink and healthy and saw it make bile. Golden bile, as
Shun said. Starzl told me I'd done a great job, told Shun he was
a spectacular teacher, and left us to finish. We lost very little
blood, finished in less time than most cases back then, and by
four o'clock in the morning when we wheeled the patient into
the ICU and I sat down to write the orders, I thought I was pretty
fucking good.

Apparently Dr. Starzl thought I was the best surgeon he'd
ever seen. At least that's what one of the residents told me later
that day.

"You should have heard him," he said. "He claimed he'd
never seen anyone operate like that. Apparently, you da man."

Apparently.

Sandee said she'd heard it, too. When Shun told me that evening how happy the boss was with the way things had gone, I
figured I was on my way. I'd become the only surgeon other than
the great Tom Starzl doing liver transplants at Pitt, and from
what I knew, we were doing more at Pitt than all the other places
in the world combined.

I hovered around the patient's bed that afternoon. I'd ordered
the routine blood tests for every four hours and could hardly
breathe when the first set was late coming back. When they did
arrive and I saw they were good, I was only happy for a short

time before I started worrying about the next set. His parents were there when he woke up and said his throat was dry. They couldn't tell enough people how grateful they were. I wanted them to know I was the one who'd pulled off this miracle, but I didn't think I should disappoint them and risk my elation.

Late that evening I realized he was going to be fine. When I went to the morgue to get my bike, I hadn't slept for at least a day and a half. I strapped on my backpack, carried the bike up the two flights of stairs, and kicked the bar to open the door onto De Soto Street.

I didn't realize it had snowed. I thought I should call my wife to come get me but remembered she'd be serving drinks in Shadyside long after midnight. The flakes were huge and floating and the sublime stillness seduced me.

I rode down the hill, barely squeezing my brakes, and took the left onto O'Hara with only a tiny wobble. By the time I passed the Concordia Club I was singing. Olivia Newton-John had been singing "Physical" on the radio when I tied the last stitch earlier that morning and I was now deeply in love with her.

O'Hara turned into Bigelow and three blocks later, Bigelow turned left and I plowed through a thin layer of wet snow like I was on rails and went straight onto Bayard. Turning onto Ellsworth, I saw a guy on a red bike.

I came down on him like a fighter pilot, and waiting till the last minute, I pulled out and swept by him like he was standing still. I was just so strong. I relaxed and that's when he came up beside me.

"Hey, man, you need to slow down!" he shouted. He was African American and his bike looked like it cost as much as my car, probably more, since my car had caught fire a month earlier and was sitting in front of my house, now under a blanket of snow, waiting for me to fix it.

I laughed and stood on the pedals and took off like I was in the race of my life. I don't know if he followed but I pretended he had and kept my head down and pedaled as hard as I could until I saw South Aiken up ahead. Entering the left turn, I stopped pedaling, put all my weight on the outside pedal, and shoved the handlebars to the right. The bike leaned left into the turn and I was almost through it when the rear tire let go and I went down so fast I didn't have time to know what had happened until he was off his bike and standing over me.

"Hey, man, you all right?"

I stood up and looked for my bike. He held on to my arm.

"You might want to wait a minute or two," he said. "Catch your bearings."

I said I was OK.

"Even so. Take a minute now and sit over here and let the wobblies pass."

I sat on the curb and was suddenly very cold and wet and tired and shaking uncontrollably. I wanted to lie down and sleep for a while before I rode the last few blocks up the hill to my house.

"Who the fuck you think you are anyway? You speak French?" he said.

"What?"

"Maybe you're like Bernard Hinault or something. You know Hinault? They call him the Badger, you know?"

"Badger?"

"Where you headed anyway?" he said.

I told him I was almost home. "Just up the hill there." I pointed up the street.

He fetched my bike from under the bushes along Aiken. "Looks OK," he said and banged the tires on the asphalt a couple of times.

I stood up and reached for my bike and felt a burning inside the icy wet on my left thigh.

"You'll have some road rash there, I reckon," he said, watching me rub my thigh.

He helped me onto my bike and pushed me a short distance to get me going.

"Take it slow, man."

I don't think I said anything.

Over the next two weeks, I felt like a rock star. Everywhere I went, people told me Starzl was raving about what a great surgeon I was.

Not bad, I thought. But I could do a lot better if I didn't have to be a contortionist most of the case. The patient did well, had no complications, left the hospital earlier than usual, and was back home attending school by the start of the next semester. I wasn't buying the best-in-the-world hype, but I thought I was really good.

I got my second chance a few weeks later. I knew it would be a harder operation from the start, but Shun kept me from getting into trouble and when Starzl got back with the liver, we were ready. I had no idea how consequential that night would become.

In the distance of more than thirty years, I can no longer claim in good conscience that I would have done a lot better had Starzl left Shun and me alone to sew the new liver in, but at the time, I was no less certain of that than of the setting of the sun. Sewing the first blood vessel together took nearly an hour, far longer than I should have taken to connect all four blood vessels and release the clamps. Most of the time I simply couldn't see the target for my needle for more than a fleeting second or two. It was like sitting in the theater behind Kareem Abdul-Jabbar with an Afro. I'd move my head one way and catch a glimpse of where to go and Starzl would shift his stance and there was the back of his head. Sometimes he'd grab my hand and try to guide the needle to the right place.

"Not there! Here!" he'd say.

"Dr. Starzl, I can't see."

"Well then get in a position so you can, for Christ's sake. If you're going to be the surgeon, you have to learn to do it."

The next day, the patient's blood tests suggested the liver had been badly damaged.

"He'll never make it," Starzl said on rounds. "We need another liver."

I walked the halls and found people looking the other way when I glanced at them. Sitting in a bathroom stall, I overheard some residents laughing about my downfall, from one of the world's best to average at best. "No, actually, below average now that I think more about it," one of them said, imitating Dr. Starzl's voice.

The patient recovered, though. I was thankful for that. Nearly ecstatic, actually. He may have taken weeks longer than most, but he got better and went home.

I was demoted immediately and didn't get a chance to do another liver transplant for months. I lobbied covertly, whining to anyone who'd listen, but mostly to Shun, that I'd been cheated, that no one could have done that transplant the way Starzl took up all the space and blocked my view and, well, didn't trust me. If he thought I needed so much help, he should never have let me do one in the first place. And so I rode the bench and did what I could to be there, to stay out of the way yet get my hands into everything.

I'd always thought the hardest part of sewing the new liver in would be the upper vena cava. It was a hard place to see, being well up under the diaphragm, especially if someone weaker and less dedicated than Hong was pulling on the rib cage. But on that first case, I'd found it surprisingly easy. I'd never done anything like it before, but I'd seen Starzl do it so many times that watching my own hands, I saw them mimicking his moves, even the way I twisted my wrist to re-grab the needle and pull it through. It was like watching him. Somehow I'd learned to do it the way he did just by watching, and watching some more.

When I finally got to do another liver the following summer, it felt like someone's afterthought. Shun caught me on rounds and told me I'd be doing the transplant.

"When?" I said.

He said we'd start around midnight, that Dr. Starzl would be back with the liver by two or three.

"He's very tired today," he said. "Maybe sick. Maybe he goes home."

That's pretty much what happened. Dr. Starzl brought the

liver back, fixed it up on the back table, handed it to me and watched us for a few minutes, then quietly disappeared.

So began my career in liver transplantation. I suppose it all seemed anticlimactic, but not to me. I felt like I'd been released from prison—put on parole, perhaps, but free to begin feeling like a surgeon again. Better than that, a liver transplant surgeon, one of the few in the world in 1982.

# PART TWO

# The Front Lines

# Hip Waders

In the early 1980s, we were still learning about all manner of things related to blood coagulation and liver transplantation. I learned later that Max Stinson's operation was one of the least bloody among the first Pittsburgh cases. Even so, for a few years, the volumes of blood lost sometimes suggested we weren't learning fast enough.

Back then, we often bathed in blood from the waist down. I knew I had to do something when I got down to my last two pairs of underwear and my shoes smelled like gangrene even after washing them in bleach.

I don't mean to exaggerate the bleeding, and we did eventually learn how to do almost any liver transplant without losing much blood, but even in those early days, we sometimes lost so little blood that as we pulled back the drapes and called for the gurney the anesthesiologists gave each other high fives. The range in the amount of bleeding I saw back then was wide, from not so bad to so bad we used stands to lift us above the pools. Sometimes when a visitor took a break he wouldn't realize how wet he was and you could follow his trail of red shoe prints all the way down the hall to the restroom and know which stall he'd used.

We tried all sorts of tricks to handle the blood. We made

troughs along the sides of the surgical drapes to hold it and keep it from spilling out on us, but the drapes were cloth and the blood seeped through anyway, and sometimes when the ventilator fired and pushed the diaphragm down, blood spilled in waves that breached the troughs. I tried going without underwear for a while, but the blood dripped off my scrotum, trickled down the inside of my thigh, and pooled in the squishy insoles of my shoes. I used an old pair of high-top Chuck Taylors for my operating room shoes. I'd had them since medical school and rarely had to wash them before, and although they may have started to fray along the edge of the rubber soles, they were still mostly white. Now they looked like I'd pulled them from a medical waste dumpster. I thought if I bleached them it would get rid of the brown and the smell but it didn't. It reminded me of the odor in the gastroenterology ward at the Cleveland VA hospital where I did my first hospital rotation as a student. I'd grown up working on pig farms and cattle feed lots and I told the VA intern this was worse than that. I said it smelled like I always thought death would smell. He told me to get used to it.

I had a pair of hip waders that I thought would solve the problem. I bought them when I lived in Utah so that I could hike the marshes of the Bear River refuge and shoot ducks. Or geese. I shot my first and last snow goose while standing in knee-deep water surrounded by icy shards and I was warm and dry inside my hip waders.

One night I was awakened by a call to come in and help Dr. Starzl. I didn't understand what was going on at first. Earlier in the day, Shun had sent me home because I was acting too goofy. I'd flown on two Learjets to two cities in Florida to get livers, helped Starzl with the transplants, and was making rounds in the ICU and delivering a nonstop string of amazingly clever one-liners when he tired of my silliness. "You're worthless to me," he said. "Go home, get some sleep."

So someone was on the phone insisting I needed to come back. She said they'd been calling and calling. I asked her what time it was and she said it didn't matter. Dr. Starzl wanted me there now.

I pedaled through a light drizzle on greasy streets and made a mess of the floor in the morgue stowing my bike. It wasn't yet two a.m. when I arrived in the operating room, and Dr. Starzl had been working for hours. It was hopeless, he said. He didn't want to quit, but he was exhausted. I was the last recourse. It was hopeless but I think he wanted to see what I could do.

The anesthesiologist told me they'd lost twenty pints of blood, or nearly twice his total blood volume just making the skin incision. I asked who the patient was and when they told me, it made sense. He was a college kid from New Jersey who'd been waiting in the ICU for weeks. His belly was a nest of vipers, huge veins radiating outward from his navel, a caput medusae larger than any I'd seen before. I worked for an hour or so on stopping the bleeding and we gradually caught up. When Starzl came back he decided to go ahead with the transplant and I remember looking for some hint that he was happy with what I'd done.

Not long after we restored blood flow to the new liver the bleeding started up again. It happens sometimes, and now we know why, and how to treat it, but we didn't back then and all we could do was keep working, stitching and packing and stitching some more. I don't remember how long Dr. Starzl stayed around and I don't remember what I said to the family in the waiting room after we finished. In the end we lost nearly a hundred units of blood, the most I have ever witnessed, and it consumed all the available blood in the city and county and maybe beyond. The patient woke up a few days later and he eventually went home.


The problem with hip waders is that they only come up to your hips. If you're walking in the Bear River refuge and you get into water that's higher than your hips, your hip waders become a cold, sloshing anchor. During surgery on the kid from New Jersey, the blood that soaked through above my hips ended up puddling in the bottom of the boots. I looked into buying a pair of chest waders but they cost too much. I told Diane about it and she showed me an apron she wore when doing autopsies. She was the pathology resident who told me I could keep my bike in the morgue, and she gave me a plastic apron that covered my chest and went down below my knees. It was clean but the plastic was clouded and yellow. She said she thought it was from the formaldehyde.

With the apron and the hip waders I was covered, but I sometimes got so hot and wringing wet inside all that rubber and plastic that I'd nearly pass out. I tried to reserve the outfit for cases I knew might be bloody, but I was a lousy prophet. I don't know when I stopped wearing the waders, but I know I threw them in with the hazardous medical waste after a long case because they had begun to leak and they stank, though not nearly as much as my Chuck Taylors. For half the price L.L.Bean was asking in their Christmas catalog, I found a pair of gum rubber boots at an army discount store, pulled the felt insteps out, and made do with the apron.

We had a resident from Gatlinburg, Tennessee, named Jefferson Davis rotate on our service. He was helping on a transplant one day when I cut into a large blood vessel and the blood shot about three feet above the table. When it hit his face, Jeff Davis leaned into the spray, growled like a bear, and shook his head like he'd just come up out of the Roaring Fork with a rainbow. I was impressed and we laughed. That was before we knew about AIDS.

A few years later the FDA approved a new test for HIV. In the winter of 1985, after I'd moved to Nebraska, Tom Starzl called and said he had some bad news and that I should sit down. He said they'd done testing on the stored blood samples of Pittsburgh liver transplant recipients.

"Some were positive," he said. "And you operated on many of them."

"Maybe they got it from the blood transfusions," I said. "Maybe we're at low risk."

"Well, we've all been tested here," he said. "I can't tell you the results, of course. All I can say is . . ." He hesitated for a few moments. "Well, my advice is that you get on a plane and go to Chicago or Denver, wherever they're not likely to know you, and get yourself tested. Unless you don't want to know. I'm not sure what you could do anyway, although you'd know whether to get your wife tested. Or your son."

In the winter of my first few months in Pittsburgh, a surgeon from Turkey named Münci and I were helping Dr. Starzl do a liver transplant on a patient with type B hepatitis. The vaccine hadn't yet been invented and we were pretty worried about getting hepatitis. Starzl said he'd already had hepatitis, in Denver. He got very sick and almost died. One of his colleagues *had* died. It's a normal risk, he said. Münci and I put on two pairs of gloves, hoping it would reduce our risk, but when Dr. Starzl was sewing the portal vein together he stuck the needle into Münci's hand. It was tight quarters in there and Münci was pulling the duodenum out of the way so Dr. Starzl could get his needle in, and each time the needle went through his glove, Münci jumped and Dr. Starzl yelled at him to hold still. You're going to rip the vein in two, he told him.

The average incubation period for type B hepatitis is two months, and sixty days later Münci turned orange and nearly died. He took months to recover and by the time he came back to work we heard there was a new experimental vaccine available. I signed up in the first group to get it.

I never got tested for HIV. I'm not sure why. I'm typically obsessed with such worries. I knew from the literature of the time that the risk for a surgeon operating on an infected patient of contracting HIV was exceedingly low—nothing like the risk with hepatitis. I thought about calling Shun to find out what had happened there, but I didn't. I later found out that when I applied for life insurance earlier that month, they'd tested me for HIV. That was before they had to get your consent. I assume I was negative. Otherwise I don't think they'd have sold me the life insurance.

# Good Opera

On the table lay a man from Kansas. He had a wife, two daughters under five, and a bad liver. He had a belly full of fluid, skin glowing like a pumpkin, and a nest of veins like snakes between my knife and his liver.

I looked into the eyes of four surgeons scrubbed and waiting to help me. Shun was in the lounge smoking. He'd helped me do dozens of transplants by then and he figured I'd call for him if I got into trouble. Hong, the Human Retractor, grinned at me. I asked for the knife and we began.

Sometimes I thought I was watching myself operate, my hands belonging to someone else. I was surprised by how closely they resembled Starzl's—the way they held the scissors, the way the index finger wiggled into nonexistent spaces between tissue planes, encircled engorged veins, and held gentle pressure against a bleeding slit. I realized this surgeon knew what to do.

As the volume of cases grew, influenced in large part by Medicare approval of the procedure, Pittsburgh became the mecca for liver transplantation. Surgeons from all over the world arrived in droves to learn how we did it.

Often, we'd have two or three cases going at once, with the boss working in one gallery, me in the other. The visitors always flocked to his gallery. On one of those occasions, I asked my sole visitor, a professor of surgery from Milan, why he wasn't over there.

"Too crowded," he said.

I wondered why that was always the case.

"You make it boring," he said. "We come to see opera. Good opera has the smell of *morte*."

Most of them took place at night—these battles we sometimes lost but always survived. It's not that we wanted to be up all night or that we could breathe easier once the sun came up; we simply had no control over when the phone rang. We went without sleep, oblivious to everything, living in a world where we knew it was morning when the hospital cafeteria was serving eggs. Time reset again and again whenever a phone rang in the night.

That's what brought Ellen Hutchinson and me together on a winter night in Pittsburgh. The phone rang at different times in different places but for the same reason. It pulled us from our beds and sent us into the night both afraid and expectant.

I'd been training in Pittsburgh for nearly two years by then. My car had caught fire that summer and the divorce had taken whatever cash I might have had, but if I was careful, even in the winter I could make it on my bike from the apartment on Marchand Street to the hospital in less than twenty minutes.

The night I went to meet Ellen Hutchinson was a Sunday. I remember that because of the smell that was in the air as I turned off O'Hara and pedaled up De Soto, a pungent mix of the odor from an electrical short circuit and the coal smoke from my grandmother's stove. It was inescapable, even in the filtered air of the operating rooms. I associate that smell most with Pitts-

burgh winters, when a featureless gray blanket blotted the sky
and grew impenetrable through the night.

I asked about it my first month in town. We were just getting
started on a transplant surgery that would take us nearly two
days to complete. Chester was the scrub tech that Sunday night.
I found him in the operating room laying out the instruments on
a large table.

"What's that smell, Chester?"

He stopped stacking instrument trays and looked at me with
his head cocked.

"Smell?" he said.

"Yeah. It's in the air at night around here," I said. "Take a
deep breath. You can even smell it in here."

"Oh, that." He opened a packet of sterile gloves and laid them
on the table. "That's the coke furnaces, Doc," he said. "You're in
the steel city now."

The nurse said they cleaned them on Sunday.

"But not till night," Chester claimed. "Not till after the
Steelers game, when everyone's too drunk to care."

Now, on this Sunday night, I rode down glassy black streets
through coke furnace smog and parked my bike in the morgue. I
was supposed to give Mrs. Ellen Hutchinson a new liver. That's
what I told her husband. He signed the piece of paper that said
he understood all the other things that could happen, but I think
we both figured it was just a formality.

"Legal stuff," he probably told his neighbors back in Aliquippa when they asked him about it later, after he buried Ellen and gave away her shoes and hats.

I'd seen people die during surgery over the years but I was never in charge then. Someone else was—someone older and more experienced, someone whose position or title or credentials, if not always his presence, gave me asylum, and alibi.

Eventually, I worked them out, the deaths in the operating room. Most of the time, I looked for something about the patient. Maybe if he'd taken better care of himself, stopped smoking or drinking so much; or if she'd come in earlier, before things got bad; or if he wasn't so old. Sometimes I told myself I was the only one who could have saved the patient, and if I couldn't do it, well, then . . .

This strategy fell apart the moment I considered the possibility that someone else, someone smarter or more experienced or just better than I, maybe even someone I knew, would have done what I should have done. Then I wouldn't have ended up with blood on my hands and soaking through to my underwear.

I have to keep in mind that no one else was available when Ellen Hutchinson's time came. The real surgeons were out of town at a meeting, off to Venice or Kyoto or Boca Raton or wherever they all went that time. I was still just a fellow in training. I had to get special permission, emergency operating privileges, to do Ellen's transplant.

The truth is I knew what I was doing. I'd operated alone more than enough to know everything would be fine, which is what I told Mr. Hutchinson when I went to see him and Ellen in the holding area before the surgery.

"How old are you?" he asked.

I told him I was thirty-three years old.

"Is Dr. Starzl here yet?"

I told him Dr. Starzl was out of town. "He travels a lot," I said.

"We came here for Dr. Starzl," he said.

I explained that Dr. Starzl had trained me. Ellen was lying on her back looking up at the ceiling while I probed her abdomen with my hand. Her liver was huge and hard and came down almost to her hip. Mr. Hutchinson leaned over so his wife could see his face.

"I've been doing most of the transplants," I said. Mr. Hutchinson wouldn't look at me. "For three, four months now."

Ellen Hutchinson stared at her husband's face without turning her head. Her eyes, yellow-stained, were sunk deep into her skull. Mr. Hutchinson brushed a tuft of frizzled gray hair off her forehead. He was missing a finger on that hand.

When I arrived in the OR, Ellen Hutchinson lay naked and asleep on the operating table. The scrub nurse stood over her tables arranging instruments, drapes, gowns, and gloves. She had her back to the room and was gowned and masked. At the head of the table, the anesthesiologist sat low on a stool filling out paperwork. It wasn't the guy I hoped would be there. He looked up when I leaned over to take a closer look at the patient.

"Well, isn't this exciting?" he said.

I nodded and looked around the room. I spotted the cart I needed in the corner of the room near the back door and went to get it. The scrub nurse looked up.

"Hey, Bud," she said.

"Sara," I said. "How are you?"

"Oh, I'm fine," she said. The corner of her eyes wrinkled and I knew she was smiling. "Question is, how are you?"

I picked up the roll of padding I needed from the cart and went to wrap the patient's legs.

"Hey," she said. I looked back over my shoulder. "It's going to be fine."

"Yeah, it'll be fine," the anesthesiologist said from across the room. "Long as you don't fuck it up."

I stopped with my back to him and rested my hands on Ellen Hutchinson's legs. I closed my eyes and listened to the background roar of the air handling system, the hiss of the anesthesia machine, and the distant beeping of the heart monitor.

I felt a hand on my shoulder and turned to see Ross standing behind me.

"Professor," I said.

He grinned behind his mask. "Mind if I join you?"

Ross was a famous surgeon from Newcastle upon Tyne. Northumberland, as he said. Like so many others from around the globe, he'd come to Pittsburgh to learn about liver transplants. Unlike most of the others, however, Ross always showed up early and he stayed to the bitter end. I looked up toward the windows ringing the ceiling. The gallery was still empty. By the time we got started, half a dozen surgeons would be up there watching, or sleeping with their faces pressed against the glass.

"Of course," I said. I looked past him at the anesthesiologist. He was on the phone laughing. "This should be a quick one."

Ross raised his shoulders in a cringe. He grabbed my wrist.

"Remember," he said and waved his finger at me. "Never tempt the gods."

I left the others to close the incision and clean up. I found Mr. Hutchinson in the waiting room by himself. He was reading *Ladies' Home Journal*. That's what they had in our waiting room, that and *Good Housekeeping*. Some volunteer—a widow, maybe—brought them from home.

He stood and waited for me to come to him. It was early morning, five or five thirty, I think. A janitor was wrestling with a floor polisher over by the vending machines.

I don't remember exactly what I said. I'm sure that in medical school they tried to teach us how to tell someone a loved one had died, as if that were the same as telling a man you'd killed his wife.

I asked him to sit down, but he didn't. He just waited in silence, and when I told him Ellen was dead he dropped to the chair and sat there for a moment, then started shaking his head and twisting the magazine into a tighter and tighter roll. I tried to sit in the chair beside him but he rose to his feet again and came toward me.

"You told me she'd be fine," he said, poking me in the sternum with his middle finger. It was his index finger that was missing.

I wanted to tell him what I thought went wrong. We didn't have the A-team for anesthesia, I'd say. I'd never had confidence in that guy and when everything started going to hell he didn't seem to have a clue what to do and by the time they called for help and our best anesthesia guy showed up . . . well, it was too late. I wanted to tell him that we pumped on her chest off and on for more than an hour, losing her, then getting her back; that when Ross, poor old Ross, stopped pumping and looked up at me from across the table with his tired gray eyes and asked if it was finished, I pushed him out of the way and pumped the heart myself and kept going and going until finally I saw that everyone was standing back staring at me and I knew then we weren't getting her back.

Mr. Hutchinson paced back and forth shaking his head, talking to himself, now and then slapping the magazine against his thigh. The floor polisher was coming closer and closer and I couldn't make out what he was saying above the noise of it. I took one step toward him thinking maybe I should touch his arm or something.

"What am I supposed to do?" he yelled at me.

The janitor was having trouble with the polisher. He'd have it bumping serenely along when suddenly it would skitter across the floor and crash into a chair or the Pepsi machine.

I asked Mr. Hutchinson if I should call someone, maybe someone in his family.

"Family?" he shouted. "Family? You want to know if I have a family?"

He put his head down and I thought he was going to start pacing again, but then there he was right in front of me and breathing hard.

"You just killed my family, son." He wasn't shouting anymore. "She was all I had. Now she's gone. Thanks to you."

He backed away and stumbled into a chair and sat down. The janitor had left the polisher in the middle of the room. He'd probably gone for help. I really wanted to finish for him. It didn't look that hard.

Mr. Hutchinson sat staring at the magazine open in his lap. I sat facing him. I leaned forward with my elbows across my knees, maybe trying to get him to look up at me but hoping he wouldn't.

"Mr. Hutchinson?"

He rolled the magazine up again and held it across his lap like a nightstick.

"We'll need the name of a funeral home," I said.

He looked up and said something just as the polisher started up again.

"I'm sorry," I said. "I couldn't hear." I pointed over my shoulder toward the noise.

"Sheffield," he said. He opened the magazine and smoothed it with his palm. "The one in Aliquippa, out on Franklin."

I left him in the chair. A different man was running the polisher now. He ran it gracefully back and forth in great sweeping arcs, moving toward Mr. Hutchinson's corner of the room.

Ellen Hutchinson's cadaver lay naked and drained on the operating table. They'd taken away the blue surgical drapes and turned off the room lights but the huge operating lights were still on her, and she looked sculpted from alabaster. I walked over and stood in the same place where I'd spent most of the night. The incision I'd made was shaped like the arms of a Mercedes hood ornament, and it was huge. I'd left the others to close it with big looping sutures. They'd been sloppy, and I regretted trusting them with someone they never knew.

Chester came in and walked to the back of the room, staring at the metal table where they'd piled all the surgery instruments in a stainless steel basket.

"Fuck," he said and shook his head. He picked up a big right-angle retractor blade with the tips of his fingers, then dropped it back into the basket. "Jesus-fuck."

Dried blood caked nearly everything in the basket, and brown streaks ran down the side of the table. Chester looked down at the blood pooled on the floor, black and shiny. He backed up suddenly, picking his feet up like someone who'd stepped in dog shit. He didn't notice Ellen Hutchinson or me.

"Where is everyone?" I asked.

He jumped. "Jesus, don't do that!" He shielded his eyes with his hand and squinted through the light. "Oh, it's you." He found some plastic gloves and pulled them on. "Fuck if I know, Doc." He grabbed hold of the instrument table with a hand on each side. "Probably another trauma. They sent me to fetch this shit is all I know."

I watched him push the instruments out the back entrance, and as the door swung shut I saw a man sitting on the floor behind it. He was asleep with his head lying on his arms across his knees. He still had his surgical mask on, and I wondered if he thought it protected him.

"Ross?" I whispered over the cadaver and across the room. I

walked over to make sure. "Hey, Ross." I had to shake him. "What are you doing?"

He looked confused at first. He was an old man, and he'd been up all night with me. He was a highly respected professor from Great Britain. He had no business sleeping on the floor.

He smiled at me.

I felt very tired.

I went back to the table. They'd washed the body. They'd also removed all the tubes from her, and I worried about that. They should have left them in for the autopsy. A white plastic sheet and a roll of flat white twine lay on the floor beneath the operating table along with a folder of papers. The Death Kit. They must have left in a hurry.

Ross was across the table from me, still with his mask up. I reached over, pulled it down, looked at his mouth, and asked him if he knew where to find a gurney. He nodded and left.

I stood beside Ellen Hutchinson's body and thought about all there was to do: tie the hands together across her body with the twine, wrap the body in the white plastic, put it on a gurney, fill out all the forms in the Death Kit, and put a name on why she died. Then cover the gurney with a white sheet and roll it down to the elevator, to the tunnels, to the morgue, and then get on my bike and ride away because someone is going to need this room to do some regular surgery, the successful kind where the patient . . . where a poor old woman from Aliquippa, the only family a man has, doesn't die because someone doesn't know what he's doing.

Ross came back and said he couldn't find a gurney. He watched me knot the string I'd looped around the hands.

"Go home," I said. He looked exhausted. "I can get this."

He came over and stood beside me and I felt his hand on my shoulder. I clenched my jaw, but tears flooded my eyes and my legs buckled. I staggered backward, away from Ellen Hutchinson. Ross tried to hold me up but we both fell to the floor.

# Believe in Life

In August 2012, my wife Rebecca's grandmother and father died within ten days of each other. A month later, my second stepmother, who was twenty-five years younger than my dad and a teetotaler, called to tell me she had leukemia. She died a few months later, four days before Christmas; my father died of a failing heart the following summer. They all had incurable problems and it seemed right for us to give up on life, and then we spent countless days at their bedsides waiting on biology to do its thing.

Marvyl was Rebecca's grandmother. In February, we celebrated Marvyl's one hundredth birthday. On the morning of the last Sunday in July, she had a stroke. The attendants found her on the floor in her room and she couldn't talk or move her right side. Rebecca said Marvyl was her usual self when she saw her on Friday. She'd found her playing bridge and too distracted to talk. Rebecca thought it was pretty precious, the way Marvyl and her friends took the game so seriously. "Oh well," she said. "I'll see you Sunday."

At the hospital, we watched the emergency room nurse fuss over the bruise on Marvyl's right arm while Marvyl clawed at Rebecca with her eyes and tried to speak. The right side of Marvyl's face was flaccid. She couldn't move her right arm or her right leg and she couldn't talk, and all I could see in her eyes was terror.

"I got an X-ray of that," the nurse said, pointing to the bruised arm. "It's OK. Nothing's broken. I had them X-ray the whole shoulder, collarbone and all, and it's fine."

She said they had Marvyl on the stroke protocol but Marvyl's blood pressure was higher than a land speed record. I wanted to say something but realized it might be better if the high blood pressure made the stroke a little worse. Or a lot worse. Let their oversight become Gram's best friend, I thought.

The doctors said it was a bad stroke. How bad, someone asked. When the doctor explained how bad, we decided to give her what they call comfort care measures only. They moved her to hospice, stopped giving her any of her medicines, and only took her blood pressure and pulse and counted her breathing twice a day, and they gave her morphine and Ativan if anyone thought she was in pain or upset.

Four days later she didn't look upset to me, not anymore. Four days later she wouldn't—or couldn't—open her eyes. I picked up her left hand, the good one, and squeezed it. She didn't squeeze back.

On Monday, she had squeezed back, hard and relentless. I thought then it might have been a mere reflex, not something intentional, but I hadn't said anything.

I saw a flashlight on the table next to her bed. I shined the light in her eyes but her pupils didn't respond. She was still breathing on her own and the bag collecting her urine was nearly full, so all that was still working.

"It's still hard to know," I said.

No one said anything. I wanted to explain that her pupils were fixed and dilated and that meant her brain was likely gone, but I didn't see how that would help.

That afternoon I told Rebecca about the line, the one you cross when you decide someone shouldn't live any longer. "We're not giving Grams any fluid or food or any of her heart medicine," I said. "Instead of giving her something that would make her die, we withhold things she can't live without."

We all wanted Grams to die in peace. We all understood she'd never be herself again, never who she'd want to be.

Someone said that Peter, Rebecca's cousin, was reluctant to go to the hospital when the rest of us were there. I worried that Peter believed we were killing Marvyl and I understood how he felt. I'd struggled with the same conflict too many times. The line is always blurred; it's made fuzzy by all the uncertainties and our emotions and by the persistence of hope. I said I could talk to Peter, share some of my experiences. An aunt said that wouldn't change his mind, but I had no intention of changing his mind. I just thought that if he was struggling he might like to know he wasn't alone.

The days dragged on and anticipation disintegrated into impatience. I found myself remembering times when the choice of giving up or fighting on in the face of seemingly impossible odds wasn't so clear. With Marvyl, not knowing when it would end was the hardest part. I thought about how I could fix that if I had a syringe with some potassium chloride. Gather round, I'd say, and poke the needle into a vein in her arm and push the fluid and wait a few seconds for her heart to stop. There. Now it's over. We can make our plans.

Of course, that would be murder.

I have been seduced by the idea of murder's deliverance before. It happened during long, horrible operations when continuing seemed a waste of time, when the life I'd been working to save had become a lost cause and it made the most sense to walk away, tell the nurses and the anesthesiologist and the wide-eyed medical student across from me, We're done. Shut down the machines, turn off the lights, let's all go home and get some rest so we can come back tomorrow and start clean.

Tom Starzl knew that seduction. He spent his life at war with it. Whenever we found ourselves late at night, half a day or more into a liver transplant, sleepless and barely able to stand and losing faith in human survival, he would feel us giving in to it. He'd be in the middle of sewing a tiny blood vessel together and the guy with the suction would be slow to clear away blood and suddenly Starzl couldn't see where to put a stitch, or Hong, faithful Hong, would begin to blink and relax his pull on the rib cage until it blocked Starzl's view.

"Shitfuckgoddamn!" Tom would yell. Sometimes he stomped his feet as in a tantrum and then he'd say it in the most piercing way, his dagger phrase that cut to the truth. "I don't want anyone here who doesn't believe in life."

I first contemplated murder in Pittsburgh. I was ten hours into a liver transplant on a guy from Durango, Colorado, when I wondered if I should remove the clamp and let him bleed out. It would be over in less than a minute and we could all go get breakfast and complain about burnt bacon. Worse things could happen to Mr. Durango. I could just tell his wife we'd done our best, and she'd probably thank me for it.

Durango had received a liver transplant more than a decade earlier. His transplanted liver had failed and now he needed a new one. When I started the surgery, I discovered everything inside his abdomen was welded together with scar tissue. Every millimeter of progress was torture. I used whatever tricks I had back then but I just couldn't stay out of arteries and veins and loops of intestine. I stopped to repair whatever I cut or burned, ripped or tore. I knew I'd never catch up if I put any of it off. After nearly five hours of that, I got a call that said the donor team was thirty minutes from landing.

Time was running out. I had to be ready to start sewing the new liver in within an hour or it would start to go bad. I had to try something else.

I worked to worm the tip of my finger around the vena cava at the top of the liver, right where it goes through the diaphragm and into the heart. I kept pushing and wiggling and then there was that wonderful, satisfying pop and my finger went through to the other side; I felt a gush of relief and asked for a clamp. But then I pulled my finger out and heard the sickening rush of blood and knew that I'd just poked my whole finger right through the wall of the biggest vein in the human body. Unless I plugged the roaring hole and stopped the rising black tide, Mr. Durango would bleed to death in less than a minute.

I don't know that any other surgeon could have done any better. I patched the hole and took the old liver out, but by then

Durango had lost a lot of blood. When the donor surgeon popped in carrying the Playmate cooler, I was pretty sure Durango had suffered brain damage from the long period of low blood pressure.

They had the new liver ready sooner than I'd hoped. We were up against a deadline, beyond which the liver had little chance of working. I wanted more time to work on the bleeding, but instead I had to sew the liver in and restore blood flow to it before time ran out.

I worked furiously, believing that once I got the new liver all hooked up, the bleeding would ease, become manageable. It didn't. Three hours after the new liver was in, three hours of burning and stitching and packing and waiting and burning and stitching some more, we were still having trouble keeping up. José, the anesthesiologist that night, said they were running out of blood. I told him to try giving more platelets and to turn the room temperature up some more. "He feels cold," I said. "Maybe too cold to clot." He looked at me and rolled his eyes and I asked him if he had any better ideas.

After eight or nine hours without rest, I packed everything off as tightly as possible with sponges and took a break. I left a resident in charge but when I came back the only person still scrubbed in was a medical student. He had two sucker tips stuck into the sponges and both were making a lot of noise. I put the retractors in, pulled out the sopping sponges, and told the medical student to suck, and that's when I realized we could win, that Durango still had a chance. All of the bleeding seemed to be coming from one place, not everyplace, and I could fix that.

The blood was streaming from behind the liver and I was a bit worried about that. I'd torn the back wall of Durango's vena cava several times while sewing it to the new liver and I'd patched it with some tendinous tissue I cut from Durango's abdominal wall. The last time I looked back there it seemed to be holding up, but I knew it could easily tear again, and that would be a disaster, so I lifted the liver very carefully.

The vena cava was good. Blood was spraying from a vein in the diaphragm. One stitch would stanch it. I asked for a suture and had the medical student pull up on the right lobe of the liver so I could get behind it. I told him to be careful, not to pull too hard. "You pull that fucking cava loose and we'll be here all goddamned week," I said. But that's what he did, right after I fixed the bleeder, and suddenly blood was roaring out of a tear in the vena cava.

Somehow, I fixed that, too. It wasn't great, but I thought it would hold. About the time I started to connect the bile duct, though, a wave of dark blood washed over the edge onto the floor and I knew it had broken through somewhere. I got everything clamped again and took a deep breath and thought maybe it was time to quit, just pull the clamp off the cava, let it go, walk out of the room, find the family in the waiting room, and wait for his mother to say, Well, at least he got his chance; that's all we can ask.

I stood with my hand on the clamp and closed my eyes. This was a broken human being. He'd spent the past three months in the hospital waiting for a new liver. He was emaciated and last week he'd grown sleepy and confused as his liver failed. I was pretty sure the chest X-ray from yesterday showed early pneumonia and I knew his kidneys had begun to shut down as we got him prepped for surgery. Even if I fixed the vena cava again, the liver was probably shot and even if it worked, there was all that recovery time to come, time waiting to die of an infection or rejection or a bile duct leak or a ruptured blood vessel, and that's only if I got him off the table. From what I could see, we were another five or six hours from that. I opened my eyes and looked at José. He rubbed his eyes and turned back to his chart without saying anything.

The medical student stepped down off his stand and bent over with his hands on his knees like he might puke. The circulating nurse asked if he needed to sit down but he said he was just a little dizzy. She got a cup of ice water and a straw; he sat down and drank it and she got him another. He was soaked with sweat and he had to change his gown and gloves. When he came back to stand opposite me he looked better, and I felt the weight of my skull on the end of my neck and the wormy wetness of blood between my toes.

I tried to remember when I'd slept last. Maybe up on 4 North, while dictating. A nurse had roused me and I saw that I'd drooled onto the chart. Now I had this clamp in my hand and I just wanted to lie down and sleep, just for a while, just until I could wake up and there'd be clear, clean light streaming through my bedroom window again and I'd know I was alive and safe and loved.

I thought about the old Victorian house I'd bought in Pittsburgh, how there came a time when I realized I didn't have enough money, would likely never have enough money to fix it up. That house was a lost cause. Let it go. Let it fall into its basement. Take off the clamp.

But I didn't let go. I'd been there before and I knew the trap.

This was when the Great and Powerful Starzl was at his most horrible best, when he kept people alert and on edge with insults and accusations, whining and yelling, stomping and kicking. "I don't want anyone here who isn't interested in life!" he'd yell. "And suck, goddamn it!" So we'd keep working, pulling and pushing on ribs and muscle, duodenum and colon, and sucking the blood so he could see and do what he had to do to save another life, and most of the time, no matter how hopeless it all seemed in those dim hours when other battlefields grew quiet and warriors found rest, we fought on and won, and notions of hopelessness were reimagined, if not obliterated.

I asked the nurse to call the resident back. He must have fallen asleep, I said.

"Are you OK?" I asked the student. He nodded, blinking.

I had him hold the liver again and by the time the resident came back I'd fixed the hole and removed the clamps and the bleeding had dropped to an annoying ooze. In all, we worked for twenty-three hours on Durango. When we began to close, he was dry and the liver had started making bile and José said Durango's kidneys were making urine.

I'm sure I talked to his family afterward and told them he'd been through an awful lot but that he was stable for now and we'd know more in the next couple of days. I don't remember any of it. Durango left the hospital a month or so later. He was still living in Durango when I left Pittsburgh. He may be yet.


Rebecca's grandmother died on Sunday morning, a week after she suffered the stroke and fell and bruised her arm but didn't break it. We were at home in bed when it happened. Worst case, I was an accessory before the fact.

Most of the rest of the family was there when we arrived. Marvyl's mouth was hanging open and I tried pushing it shut but the muscles were already stiff, so I held her chin cupped in my hand for a long time while the others told stories and laughed and cried and made the plans. I thought her mouth looked better when I let go, though it still hung open a gap. I figured the undertaker would fix that later but then I remembered Marvyl had wanted to be cremated.

# Concussion

When my sister was seven and I was nine, she ran headlong into a clothesline post thick as a telephone pole. We were at our grandparents' farm outside Danville and she was running away from Scout, their dog.

The blow knocked her out. I heard my mother screaming on the front porch and turned to see her leap from the swing. My sister was bawling pretty loudly by the time Mom got to her. Dad came around the house from out back. I'd never seen him run before. He dropped down beside her and patted her on the head, rubbed her arm a few strokes, looked at my mother.

"The girl will be fine," he said.

Then he looked at me. Had I been chasing her? Had I thrown the baseball at her? (I had, but that was earlier.)

"What the Sam Hill have you done now?" he said.

I mentioned Scout. Dad stood and looked around. He found a stick and whistled for the dog.

"I thought she was dead," I said. "But she was only knocked out, I guess."

I didn't want Scout to get a beating. Dad squatted down right in front of my face and suddenly I wished I'd kept quiet.

"What are you talking about?" he said. He threw the stick into the scrub along the fence line.

I told him how she didn't move for a few seconds. He asked me if she'd done anything else.

"Did she start twitching or making jerking motions? Like this?" He held his arms in front of him, flexed at the elbows, and made some rigid shaking motions and I couldn't think.

"Well, did she?"

"I don't know," I said. "I might have looked away for a few seconds. When Mom screamed."

"But you didn't see her do that?" he said.

I felt like crying. I didn't know I was supposed to look for that. I should have watched more closely.

They put her in Grandma's bed. Scout came in and lay on the rug by the dresser. Dad said not to give my sister anything to eat or drink. She fell asleep. I thought that was wrong.

"Shouldn't we try to keep her awake?" I said.

Dad laughed and I wondered if he really knew anything about it.


Later that morning, Dad took a .22 rifle from the hall closet and asked me to follow him out back. Scout was still in the bedroom and my sister was still asleep.

Dad pointed to a tree on the other side of the creek. "See that branch?" he said. "The skinny one just above the big flat one?"

I nodded.

"Keep an eye on it."

He rested the rifle on top of the fat wooden fence post and squinted down the sight.

"OK, you still watching?" he said.

I nodded.

"Well, are you?"

"Yes," I said.

A loud bang made me jump, and I saw the skinny branch bend in the middle, hang for a second, then fall to the ground. I couldn't believe it. He looked at me and laughed.

"Not bad, huh?" He looked back at the tree. "Not bad at all," he muttered.

He asked me if I wanted to try. I wasn't sure what he meant.

"No big deal if you don't," he said.

"I guess," I said.

"Well, first things first," he said. "If you're going to shoot a gun, you first have to know how not to kill yourself or someone else. *Capisce?*"

I got my first lesson on gun safety. He showed me how to carry the gun, even when it's empty. He showed me the safety switch, how the bolt action worked, how to load a shell, how it ejected when you pulled the bolt back. And he told me over and over never to point a gun at anyone or anything I didn't intend to kill, whether the gun was loaded or not. I wondered if I was going to kill something.

He handed me the rifle, then gave me a bullet. It seemed so small.

"Go ahead, load it," he said.

I put the shell into the hole, slid the bolt home, and levered it down. I made sure the safety was still on and kept the end pointed down toward the creek like he'd said. He nodded and helped me lay the barrel on top of the post and showed me how to line up the sights.

"You ready?" he said.

I nodded.

"It won't kick enough to notice, but keep it in tight against your shoulder, OK?"

That's when Grandpa yelled from down by the garden. "Come quick!" he said. He sounded mad about something.

Dad showed me again how to unload the gun. He pocketed the shells, handed me the gun, and told me to put it away in the hall closet.

"Maybe we'll have a little time after dinner," he said. "Before we hit the road."

I put the gun away and checked on my sister. Scout was gone but my sister hadn't moved. I watched for a while and made sure she was breathing.

I heard Grandpa yelling out back again and went to see. He and Dad were chasing the hens up and down the path. Grandpa was holding a butcher knife. He seemed furious but Dad looked at me and winked. Grandpa handed me the knife and pointed to the workbench. I took it and laid it there alongside two smaller knives with brown-spattered handles.

A hen zigged to get away from Grandpa and ran right into my leg, then Dad had her by the neck. He was giggling but Grandpa was very serious.

"We need one more," he said and chased a clutch toward the outhouse.

Dad was still laughing when he pulled the head off the chicken and dropped the bird to the ground. The headless hen flipped and flapped and jumped around all over the yard, blood spraying out of the neck stump. It came toward me. Grandpa's foot came down out of nowhere and pinned the thing to the ground. It was still fluttering and wriggling as he pulled the head off the other hen and let it go to ground and the dance started all over again. He looked at Dad, then threw the head over the fence into the creek below.

I didn't understand how a chicken—how anything—could still do all that with its head missing. I thought of the pictures of a guillotine our history teacher had shown us, and how they used it to cut off that French woman's head and how the head fell into a basket, and I wondered if she'd flopped around like that afterward or if she could still see, and that's when I threw up all over my jeans. Grandpa reached down and picked up the hen under his foot, held it out by the legs, and started pulling out feathers. Dad caught the other one up and did the same. Feathers floated everywhere, landing in my hair, sticking to the puke on my jeans and to the blood splatters on Dad's boots.

When they were done, Grandpa laid his pale, wrinkled carcass of a chicken on the bench and cut open its stomach with one of the smaller knives. He reached in and pulled out a handful of guts and threw them on the ground next to the cats that had come out of nowhere. A gray tabby sprung on the mess and a fight broke out. They were yowling and shrieking and then the tabby ran off into the weeds, chicken guts snaking along behind. Grandpa pulled out brownish-red lumps, which he laid out on the bench. He split one of them with his knife and it seemed to have some dirty-looking stuff inside. He turned it inside out and pulled off a shiny gray layer. Some grainy matter pattered on the wood bench. He laid the reddish part next to the other bits. Dad went over and picked up the thing, took the shiny gray bit from

Grandpa, and pinched some of the grainy stuff. He showed it to me in the palm of his hand.

"The gizzard," he said. "See the little bits of gravel mixed with the ground corn?"

I nodded.

"And this shiny layer? This other part is a thick muscle and the membrane here is the inner lining. It holds the gravel."

It made no sense. I felt him looking at me.

"Chicken teeth," he laughed. "Ever heard of a chicken with teeth? No? Well this is what he has instead. The gravel stays in here, the grain the chicken eats comes down the pipe and into the sac this membrane forms, and then this really strong muscle grinds it all back and forth till the grain is a mash that slides out here and on down."

I touched the grainy stuff and rolled the tiny gravel across his palm.

"*Capisce?*" he said.

I nodded. He put the gizzard back on the bench and tossed the membrane at cats that had come back from battle, then brushed his hands against each other. I watched the grain fall to the ground.

Out of the corner of my eye I saw Grandpa's hand rise above his head, then come down fast with a loud bang. I looked up in time to see a chicken foot fly off into the weeds. He whacked the other off with the butcher knife.

"Here, boy," Grandpa held out a small towel. "Take this down to the spring there and clean yourself up for dinner."

I took the towel and looked down toward the creek.

"Get along, now," he said. "We'll be eating soon enough."

I knelt beside the spring and blew at the surface of the water to make a clearing in the duckweed, where I dipped the towel. The water in the spring was colder than I remembered from the summer before. I looked downstream of the spring and imagined chicken heads floating there among the cobwebs and tadpoles.

After dinner, I found Dad and Mom on the porch swing. He had his arm around her shoulder. They'd taken a plate of food into the bedroom for my sister, but it was now sitting on Mom's lap, the food untouched. Mom's eyes were red and Dad looked worried. I'd never seen him worry before. I just wanted him to make my sister better. He needed to wake her up but my sister was sleeping so hard. I was suddenly very frightened. I wanted to ask Dad if she was going to die. I wanted him to tell me she was going to be fine, like he had told Mom before.

On the way home to southern Ohio, we stopped in downtown Mount Vernon. Dad carried my sister in his arms into a white clapboard house with a sign outside that read DR. DRAKE. Mom stayed with my little brother and me in the station wagon. I asked her what they would do. She said Dr. Drake was a very good doctor.

"He'll know what to do," she said. She shook her head and covered her mouth and looked out the window. Then she turned around and smiled at me. "Dr. Drake's the one who got your daddy to become a doctor. He took care of him when he was sick."

When Dad came back with my sister, she was awake. He put her on the front seat and she laid her head on Mom's lap.

"I'm hungry," she said. "Can we stop for ice cream?"

Mom laughed and blew her nose into a wad of tissue.

Dad said Dr. Drake gave her a shot. A shot was a big deal. "She's going to be fine," he said.

He put his hand on Mom's arm and she started to cry and laugh at the same time. I'd never seen her cry before.

# The Bell Rule

Mom is ringing her stupid bell again. All the windows in the house are open and that big attic fan is roaring away in the hall just outside her bedroom, so I know she can't hear us, which gives me a perfect alibi, since it's my turn at bat.

Teddy's a lousy pitcher and the others know it. Jim has moved back in left field. I do the Babe Ruth thing and point toward the maple tree in deep center. Ted throws a wild pitch and I swing anyway, barely nicking the ball, but it rolls toward Deebs at second, and with no one playing first we're in a race.

Deebs is so fat. Halfway to first I'm thinking I can make it to second before he gets close enough to tag me and by the time he figures that out I can be on my way to third. I round the boat cushion that is first base and Deebs is running like a duck, his head down, clueless that first base is already a distant memory for me. But then Teddy calls *infield* and I can hear Mom ringing the bell again.

"That's flies, not grounders!" I yell. I'm between second and third and it's looking like I might make it all the way home, which is probably why Teddy, who everyone knows can't pitch worth a shit, calls *infield*.

"Is not," Jim says, and I notice him motioning to Deebs, who looks like it might be time to sit down and rest.

"Is, too," I say. "It's called the infield fly rule, idiot."

Mom rings the bell again but I figure she can wait. Probably wants more ice in her Coke.

Deebs stops running and I think he's going to throw the ball. He's going to throw it to Jim and Jim isn't fat. Jim's fast and I'm trying to figure out if he's going to go to my right toward third or block my way back to second and then there it is again, she's ringing her fucking bell and Teddy's standing on third and Jim's caught the ball near second and I can see now that if they pull this rundown off, the rule won't matter.

"Time out!" I yell and make a T with my hands.

Jim rushes over and tags me with the ball. "You're outta there!" he says.

"I called time," I say.

"You can't call time out," he says. "Not right in the middle of a play."

I remind them of the bell rule. I can call time out whenever my mom rings her bell.

"She could be dying," I say.

"Yeah, right," Teddy says. "Like that time you hit the easy-peasy pop-up and called time out. I didn't hear no bell."

"Me neither," Deebs says.

"Give me a break," I say just as she rings the bell again, harder and longer than before.


My mother smoked too many cigarettes and now she has lung cancer. We found out the day Dad took us up to Columbus to see her. She was in the hospital because they'd opened up her chest and taken out a lump. They said it looked like cancer. They were hoping they could cut it all out but there was just too much. That was right before I turned twelve and now I'm thirteen and a half so I'm pretty sure the cobalt they gave her did the trick. At least for a while.

Dad says the cancer's back but they can't give her any more of that. He says it would be like if she'd been there in Japan when they dropped the bomb and all those people got sick and died. I know what he means. It's the same reason the Bumgartners built a really cool fallout shelter in their backyard, so the radiation couldn't get them.

These days, Mom spends most of her time in bed. Now that it's summer, my sister and I are supposed to take turns seeing what she wants when she rings the bell. My brother's only nine, so he's off the hook. My sister always has somewhere to be, like 4-H camp or Bluebirds meetings, where they make stupid pot holders out of towels or napkins out of sheets.

Sometimes when Mom rings the bell all she wants is another Coke or more ice in the one she's drinking. She likes to let it sit for a while so it doesn't burn so much going down, but the ice melts and no one likes hot Coke. Other times she wants help getting to the bathroom, but not so much now that they got her that pan, which is worse because most of the time she doesn't get it where it needs to be and it makes a mess. Once, when it was really full, I spilled a bunch of it on the carpet. That made her pretty mad.

She tells me I don't have to clean her up and that's fine with me, but I hate to think about her lying there all that time until Dad comes home. Thelma does it on the days she comes, which seems to be more and more now. Thelma's a black maid and she does the cooking. Sometimes she makes empanadas. That's my favorite. One time she cooked pigeons for us. She called it squab. I shot the pigeons with my pellet gun down at the staircase factory. It's closed down now and there are pigeons everywhere just waiting to be shot. I cleaned them how Dad taught me with pheasants and she fried them the same way Mom does chicken.

Mom's pissed I took so long.

"Where have you been?" she says. "I rang and rang and now look what's happened."

She's naked from the waist down and holding herself up against the bathroom door and looking down at the floor. I don't want to look but I can't help it. Her hip bones are like wings on a pterodactyl, angling down and joining in the middle to a mound of thin yellow hair. I don't believe for a second this is Mom anymore, but some diseased creature living inside her.

"Oh God," she says. "Who's going to clean this up, this mess, this horrible mess . . . ?"

She starts crying because there's a puddle like chicken gravy on the carpet and more of it dripping down between her legs onto the bathroom floor. It's linoleum and it's blue like the carpet, like a robin's egg.

I don't know what to do. She looks like she might fall down any second but it's the mess I think about, and I run down the hall to our bathroom and look for a towel but I can't ruin my Lone Ranger towel so I grab the pink one with white birds and by the time I get back she's already sitting on the edge of the bed and her feet are soiled from the puddle and all I can do is stare and wonder why Thelma isn't here, why she doesn't come every day, why I'm the one standing here hating what Mom has become, terrified by the sharp-angled bones bracing her yellow skin like a rubber tent.

I try to clean her feet but it smears, and she cries out and yells at me to be careful of her bones, they hurt so much.

"Wet it!" she screams. "In the sink, get it wet."

But when I come back and try again, it's too cold.

"Oh God, oh God, oh God," she says. "It's freezing. Make it warm, make it stop hurting. Oh God."

I rush to the bathroom and wait for the water to warm and I see her tip over to lie on the edge of the bed with her feet still on the rug. Her eyes are closed and I wish she'd put the pillow under her head because her neck is bent at such a terrible angle.

Dad comes in then.

"What the Sam Hill?" he says. He's dressed in the green pajamas he wears to operate on people. He never wears those at home unless he's in a hurry, like when I nearly cut the tip of my thumb off carving my Pinewood Derby car.

"What are you doing?" he says. "How did this happen?"

He puts the pillow under her head and she opens her eyes and starts bawling like I've never seen before. Dad grabs the wet towel from me, cleans her feet, lifts her legs into bed like they're made of glass, and pulls the sheets over her, then unfolds the quilt up to her neck.

"What a mess," he says, and he shakes his head at the carpet and the bathroom floor. "Jesus."

He's mad at me because I didn't come when she first started ringing. She must have called him.

"I had to leave right in the middle of my OR schedule," he says. "Now look what you've done. This is what happens when you don't care about your mother."

I try to look like I'm sorry, but I'm just so relieved he's here now. He tells me to go get the carpet cleaner stuff and a brush from under the kitchen sink, and somehow together we get the mess cleaned up, and Mom falls asleep.

Dad says I need to pay more attention. "It's only for a few hours. You can spare a few hours in your busy schedule to help your mother, can't you?"

Once he's gone I go back outside but they're gone. I walk out into the yard so I can see Teddy's driveway. Their bikes aren't there and I know they've gone to the pool.


One time I got to stay up and watch the *Danny Thomas Show* on channel 10. It was Dad's idea and we sat beside each other on the couch. Mom was lying in bed, just on the other side of the wall behind my head, and pretty soon I heard her calling out. It sounded like a kitten crying for water, and Dad got up to see what she wanted.

Their voices were muffled and I couldn't tell what they were saying until Mom started crying. "What about my babies?" she said. "Who will take care of my babies when I'm gone?" She kept saying this over and over again. Dad's voice was soft and low, but it didn't matter. "I don't want to leave my babies," she said. "Who'll take care of my babies? Please, God, don't make me leave them."

# Death 5, Mrs. Rothstein 1

During my third year in training at Utah, I got to spend August working in the hospital in Yellowstone National Park. I was supervised by a senior faculty surgeon from Utah. We did minor surgery on park employees—hernias, hemorrhoids, broken bones, lacerations, moles. We also manned an emergency room, such as it was, and rotated night call. Lots of gonorrhea among the staff that year. Maybe every year; I never asked. The emergency medicine doctor from San Francisco who ran the place taught me a really cool way to take out fishhooks. Fishhooks in scalps, fingers, hands, arms, legs, faces, even an eyelid. I never knew fishhooks were such a scourge on humanity.

The on-call person slept in a closet of a bedroom behind the hospital cafeteria. I spent my time before bed in the lounge, reading Vonnegut or, less often, a textbook called *Surgery*.

One evening after dinner, when the others were settled in their cabins and the kitchen staff had doused the lights and caught the park bus, I sat reading about ice-nine and Bokonon when the front bell rang and I met Mrs. Rothstein from Queens.

"I've come to see the doctor," she said. "Get me the doctor, young man."

The tall young woman holding on to Mrs. Rothstein's arm looked at me and lifted her eyebrows and I knew she'd been attending to Mrs. Rothstein for more than this little adventure. I explained to them that I was indeed the doctor. Doctor Shaw, I said and put out my hand. Mrs. Rothstein shooed me with one hand and guarded her heart with the other, a large sequined bag hanging from her elbow.

"The real doctor," she said. "I want the real one."

My hand was still out.

"No offense, young man," she said, "but you can't be over, what? Twenty years old?"

I told her I was almost thirty and a real doctor. "I'm from the University of Utah," I said. "Department of Surgery."

She looked at the young woman then, her eyes watery and wide.

"Why don't we find a seat in one of those rooms over there so this nice doctor can help you, Madeleine?" she said. She pointed to the exam rooms with her chin and leaned a little into the old woman to get her started.

Once we were seated in the exam room, the young woman left us alone, miming to me with her fingers that she was going out for a smoke. She eased the door shut and Mrs. Rothstein and I were alone at last.

"So, how can I help you?" I said.

"Where is this place?" she said. "I know we're in that park with the geysers and all, but all I've seen all day are trees and trees and more trees. We rode that bus for hours and hours through a tunnel of trees and now here we are, no man's land, a wasteland of trees and so far from anything."

I tried orienting her to the geography of the park.

"But where would they take me if something goes wrong?" she said.

"Wrong?"

"Yes. Like I fall and break my hip. Or my heart. I have a weak heart. Dr. Silverman prescribes me heart medication. What if the medicines stop working out here in the middle of God knows where?"

I asked her about the pills and she opened her glittering black bag and hauled out a sack full of pill bottles. I began lining them up on the table while she talked.

Madeleine Rothstein was a widow from Queens whose late husband, God bless him, had worked for the borough as an inspector and died of a heart attack coming up from a basement on Utopia Parkway and Kildare.

Five bottles were various vitamins or supplements. One was a mild water pill.

"We had a lovely home in Forest Hills," she said. Now she lives in an apartment near Queens Hospital. "I can go to my doctor anytime I want. I see his office from my bedroom window," she said. "Very convenient. And so safe."

She had both Librium and Valium in separate bottles, one labeled by a shaky hand, "mild nerves"; the other, "STRONG. Take only when urgent."

"Do you take these?" I said, holding up the STRONG ones.

She squinted to read the label then waved them away. "Never," she said. "Too much. Puts me to sleep for days. Wet my panties and ruined the couch, I was so out of it."

"What about these?" I held out a full bottle of digoxin, a pill for heart failure.

"Oh, no, those were Saul's," she said. "I carry them just in case."

I narrowed my eyes, trying to be stern.

"In case my heart starts to give out," she said. "God knows *you* can't help me."

I began to see what was wrong with Mrs. Rothstein and I wondered if, in fact, I could help her. I thought about what a perfect ending this was to a day that should have horrified me, and in a moment of surprising insensitivity, I wondered whether she'd like to take a look at the five bodies we had stacked in the basement.

The bad shit started just before first light that morning when the nurse manning the emergency phone woke me up over the radio. "Man down in the cabins," she said. "Sounds cardiac."

I decided the guy who'd numbered the cabins over by the lodge wasn't nearly as funny as he'd thought he was. When I finally found No. 33 (next to 27 and across from 12), I thought that a just god would have put that funny man down on the floor of No. 33 and not the white-haired, big-bellied gentleman I found lying lifeless on his back in a pool of shit and piss and dressed in a pair of red-and-white-striped pajama bottoms. Another, younger man pacing back and forth and wringing his hands asked me what took so fucking long and I apologized, set the crash kit down, pulled out a stethoscope to try and hear something, felt his wrist and neck, and decided to go through the motions anyway. I managed to insert a tube through his mouth into his trachea and persuaded the other guy to squeeze the bag while I pumped up and down on the old man's chest. Twenty pumps and I'd stop and listen, shake my head, and start again. I figured three rounds would be enough to show we gave it our all, but my friend squeezing the bag wanted to keep going, so we did, for a while, until I finally stopped and watched him squeezing until he, too, understood we were done. I reported out to the ER doc, who was eating breakfast in the hospital cafeteria. Probably dead for twenty, thirty minutes by the time I got there, I said. Took off as soon as I got the call. Had some trouble finding it, but even so, nothing I could have done.

An hour or so before lunch we got a radio call about a crash down at West Thumb. A head-on. Six kids in a car, an RV. One dead, helicopter coming for two others. Expect three by ambulance, cuts and bruises. We got two bodies, not one, and three kids scared shitless, but with minor injuries. They sent me to pronounce the two in the morgue, which was a basement room where it never got above forty-five degrees. I found three body bags on the concrete floor and the first one I unzipped was the old guy from No. 33.

A thunderstorm hit us after three that afternoon. I walked down to the lake and watched it roll in, rippling the surface at first, then lifting up white-capped waves tall as barstools. It didn't last long and I waited it out in a picnic shelter. Back in the hospital, I heard we'd be getting two more bodies. Boy Scouts out in canoes, flipped over far from shore, lake water still cold as ice melt. I had to pronounce them as well. They were still wet, dressed in scout uniforms. One had curly red hair and freckles. The skin on his forehead was so taut, and white. I zipped him away and stood looking at the five bags, wondering at my numbness, my lack of panic. I had nothing to do with these tragedies. There was nothing I could have done.

\*    \*    \*

By the time I sat opposite the very anxious Madeleine Rothstein in a tiny clinic room in Yellowstone Park I was already routinely obsessed with worst-case scenarios. It was part of being neurotic, of being born into a family of worriers, and it made me a better doctor. Worrying about what the worst thing might be could be good for my patients. That was certainly so later, when I became a transplant surgeon. Not thinking first about the worst possibilities could mean not acting in time.

I wasn't there yet when Saul Rothstein's widow found herself at my mercy, but I was already smart enough to know what was wrong with her without doing a bunch of tests.

I listened to Mrs. Rothstein's heart and lungs, felt her belly and the pulses in her neck and arms, then stood and held out my hand.

"Would you come on a little walk with me?" I said. I wanted to show her the hospital. I thought if she saw how prepared we were for just about any emergency short of nuclear attack, she could go back to her room in the lodge with that fine young woman out smoking in the loading dock and sleep the night away without another thought of bad hearts or her doctor two thousand miles away at Queens Hospital.

The operating rooms, both of them, gleamed with splendid sterility when I flipped on the lights. "We can do about anything here, except maybe a heart transplant!" I said. I laughed but I felt her shudder through her grip on my arm. I led her over to the glass cabinets and showed her some of the spare instruments. "Of course, the ones we use all the time are already sterilized and wrapped in those green packs over there," I said. "Ready for the morning cases."

The lab was just as impressive, what with the modern blood test machines and centrifuges. I opened the refrigerator and showed her the bags of fresh blood. "Ned can type and cross a person in twenty minutes and we've got every blood type," I said. "Except B. And AB. For that we'd just use O. I think. They're both pretty rare, you know."

I felt her get heavy and I had her sit in a chair for a few moments. She didn't look so good. I counted her breaths and felt her pulse and both were a little fast.

"I don't know what's wrong," she said. "I'm so faint. Can't catch my breath."

I realized she was hyperventilating and wondered if I could find a paper bag in the kitchen.

I heard Mrs. Rothstein's attendant calling from out front and I shouted back. She came running, took one look at Madeleine and asked for the old woman's bag. She fished a small pink pill out of the "mild nerves" bottle and broke it in half.

"Can we have some water?" she said. I realized she was talking to me and I ran to the kitchen for a glass. Mrs. Rothstein took her half of the pill and washed it down, and within minutes her breathing slowed and she stopped tapping her feet and working her hands open and closed. I looked at the young woman. She smiled at me, then stuck out her tongue and placed the other half of the pill on it, slowly pulled it into her mouth, and swallowed hard.

# The Weight of My Bones

I'm obsessed with the notion that I got lymphoma because I went without sleep for so much of my life. When I was twelve, my dad asked me to volunteer at a pancake breakfast to raise funds for the American Cancer Society. My mother had lung cancer because she smoked too much, and he told me the Cancer Society was trying to find a cure for cancer. He said I would have to stay up all night building tables and mixing Bisquick and filling water coolers in order to be ready for the farmers who would show up before dawn. I worked all night and I thought my dad would come by on his way to the hospital around seven o'clock, but he didn't. I sat down to rest for just a minute and fell asleep, and by the time they woke me up he'd come and gone. In college I studied so long without sleep that sometimes I couldn't tell if I was dreaming or hallucinating. A decade later, I saw my first liver transplant and Tom Starzl said I was too weak to be a transplant surgeon; in the years that followed, I set and reset personal records for consecutive days, not hours, without sleep.

Sleep deprivation was just part of being on the front lines in Pittsburgh. We slept when we could between operating and making rounds and operating again. We went home whenever we realized we were done and no one was looking for us, or when we thought we had at least a couple of hours free.

I remember a Saturday morning in the spring of '82 when I woke up to a roaring that rattled the window. I pulled back the curtain and saw a balloon floating not a hundred feet above my bedroom window. It was mostly pink and the man in the basket had a beard and looked a little worried. He stood with his hand on some lever, looking toward the hills to the west while the flames roared over his head and a blond-haired boy peered over the edge and pointed at me. It was ten o'clock. I'd slept for twelve hours.

I felt good—euphoric, even. I'd done three liver transplants and three or four kidneys and saved Mr. Hardin without more than short naps here and there over a three-day stretch.

I put one of the livers in a kid from Utica whose first liver transplant didn't work. I'd been sure he wouldn't live long enough for a second chance.

Mr. Hardin had collapsed in the hallway shaking my hand and dressed to go home with his new liver. He was thin and light; he fell toward me and I laid him on the floor. I couldn't feel a pulse, so I ripped open his shirt, wiggled my fingers into his unhealed incision, and felt the rush from a ruptured artery. I held pressure against the flow until we got him to the operating room, where I fixed it.

Now the sun was shining for the first time in a five-month Pittsburgh gray season and a ridiculous pink balloon was lost in our little ghetto. I wanted desperately to fuck my wife.

I slid my hand down the flat of her belly and gently touched her and listened to her breathing to know when she woke. She sighed, grabbed my wrist, and pushed me away, pulling the sheet up as if to sleep some more. I waited but it didn't go away and I slid close against her back so she could feel me. Her arm was a strap of muscle over bone and her elbow dug into me as I slipped my hand under her breast.

"Don't," she said.

"Really?" I said.

"I'm tired."

I rolled onto my back. The sky had never been so blue. I thought of my pager pocketed in my scrubs on the floor and felt the brief panic that comes each time it goes off. Carol and I had become friendly strangers by then and I'd taken to masturbating in the locker room shower. I couldn't help it. The urgency of it was strongest after the cases when someone came close to dying. I felt a kind of wildness that wasn't so much victory as escape, thrilling escape from near death. I was a beast and I daily walked up to the brink and jumped in, ripping and tearing and slashing and scream-ing, always crawling out the other side reaping air and worship-ping the weight of my bones. I was alive, I was well, I was strong.

I rolled into her, pulled her to me, and kissed the back of her neck. "I love you," I said. "I need you."

She leaned away and pulled at my arms. She slid out of bed, went into the bathroom, and closed the door. The clock radio made a clicking sound and I remembered lunch with Roger and Hector and Eduardo. The volume was full and Led Zeppelin came in halfway through "All My Love" and the dogs came run-ning and barking. I lay wallowing in the wonderful hard chaos until Carol came out and pounded the clock and invited the dogs to breakfast.

LAST NIGHT IN THE OR    165

"I can't take this," I said.

She stared at me for a moment but I knew she couldn't speak. Shadow barked at her and Bob clawed at her legs, tearing the skin, and she winced and fled the room.

The high of a pink balloon morning felt honest and deserved. We, those of us on the front lines in Pittsburgh in those days, often felt ecstatic just to be alive. Or I did. Maybe it was living with so much death. Most patients lived, of course, but enough of them died to remind me every waking day of my own mortality. I remember crying the first time I saw my sister's newborn son. Every baby I'd seen since arriving in Pittsburgh was sickly green with jaundice, emaciated and alarmingly fragile, and here was a child so healthy, so alive and happy, and pink. So unbroken.

Two years before we moved from Utah to Pittsburgh, Carol told me to leave her alone on a sunny Saturday morning after four months of celibacy and every other night on call. I slammed the door on my way out and drove to the ER to take Eric's shift so he could go skiing. Carol wasn't home when I got back the next morning or the one after that. When she did come back we agreed to separate but not until after I was through with my training. I guess because it was easier that way. A year later she had a miscarriage and neither of us had known she was pregnant. I guess we just couldn't talk about it. Before long we were passing through Peoria on our way to Pittsburgh in a poorly packed Ryder truck.

In Pittsburgh, we moved into a Victorian-style three-story on Harriet Street. We'd written a check for a down payment in July and by December we were still waiting for the bank to decide whether to loan us the money. The house was falling apart, at least on the first floor, where the ceilings hung down in great sheets and slat-filled holes in the plastered walls were shaped like beasts of the Serengeti. We managed a kitchen with a gas stove and running water and lived on the second and third floors. We soon found it cost too much to heat the place with the molten old furnace, so we bought portable heaters and kept them topped up with kerosene from red jerricans we stored in the basement. Someone a few blocks over was found dead in his bed around Thanksgiving and they said it was carbon monoxide from a kerosene heater. After that we bought more blankets from the Goodwill store and burned less kerosene.

The pink balloon floated away toward the Allegheny Valley. I took the dogs for a walk and watched them crap in the alley. I made coffee and found a cable TV brochure on the kitchen table. I thought we'd agreed we couldn't afford cable.

I ate lunch with Hector and Eduardo at T&K's Soul Food Kitchen on Craig Street and afterward drove to the hospital because Chris was on duty. Chris worked on 4 North, where we kept kidney transplant patients. I met her my first day in Pittsburgh. I'd been spending lots of time on 4 North. I figured out where they kept the duty roster and I tried to memorize Chris's schedule. That winter, we had a patient named Devon from West Virginia. She was fourteen years old, looked ten, and acted twenty-nine. Her third kidney transplant had failed and we were working to clear up an infection so she could try again, and several times

Chris came in on her day off and we took Devon out to lunch, even though it was against hospital rules. Lately I'd been suggesting to Chris that we run off to Aruba together. I didn't know anything about Aruba except what I'd seen in a TV ad for an airline, but I liked the tropics.

On Saint Patrick's Day, Chris invited me to her apartment to drink green Scotch with her, but Starzl found a donor and I had to operate that night instead. She said she'd save the Scotch; I went to her apartment three nights later, on a Saturday. On Sunday morning I got paged right in the middle of trying to convince her to put the anchovies from the leftover pizza on her eggs. That week we stayed together every other night.

On the next Saturday I was watching *The Wizard of Oz* on cable when Carol came home near midnight and I told her I was leaving. I drove to the hospital and stayed in the call room. I moved in with Chris the next day and on Monday, Henry Fonda and Katharine Hepburn won Oscars.

I felt so reckless, so impulsive, but it had been a long time coming. Carol and I'd failed years before. She'd put up with my long work hours and selfishness and I with sexlessness. Our resentment was neither overt nor consensual, but it was mutual. I fell in love with Chris because she loved me. Living on the front lines, trying to save everyone while death seemed so close by, not in the shadows but right there in those brightly lit rooms where we worked, I was desperate to love and be loved. At the time, I felt no guilt. I felt release. I felt alive again, and suddenly death seemed so much less inevitable.

I thought about the unpredictability of battle on a war front and how terrifying that must be compared to what we faced. I had no uncertainty about when I would die. As far as I was concerned, I was immortal. Mortals were those yellow people waiting in beds, yearning for us to save them. Only they faced an uncertainty worthy of battlefield soldiers.

In the summer of 1984, I went to Minneapolis to present three papers I'd written related to liver transplantation at an international conference. I arrived not having slept for maybe sixty hours and got a little drunk the evening of the gala dinner. My friends put me to bed around midnight to keep me from embarrassing myself any more than I already had. I slept for fifteen hours, missed breakfast, lunch, and my flight home. A month later, on our first anniversary, Chris and I flew to Nantucket to share a rented beach cottage with my brother and his wife and their two-year-old daughter. We spent about four days with them, and except for a few hours after meals, I slept the entire time.

A week or so after we returned home, I woke up about three a.m. with the worst headache of my life. I was certain I had ruptured an aneurysm. I gave my wife the donor card in my wallet and asked her to take me to the emergency room. They did a CAT scan of my head and a neurosurgeon named Bill said my brain was a little swollen. He did a spinal tap and said it looked like I had a viral infection. Encephalitis, he said. He said I should stay in the hospital. I asked what they would do in the hospital and he said not much, so I went home. I spent the next three weeks more or less in bed, waiting for my head to stop hurting. No one ever told me what kind of virus it was and back then there weren't any drugs for treating viruses, so it didn't really matter.

During my month off, Dr. Starzl and the others didn't simply pick up the slack; they nearly doubled the number of liver transplants we were doing. I couldn't imagine how that was possible. I worried he was using donor livers that I would have rejected as damaged, and I wondered who was taking care of all those patients without my guidance. I called Shun several times. Don't worry, he said. Just get better.

The headaches resolved and I went back to work but I still didn't feel right. I'd have days when I felt like all the nerve endings in my body were raw. I told Shun it was as though the gain knob on my nervous system was turned up to eleven.

Starzl had made changes in the way the team worked while I was gone, and I felt that everything I'd worked to make better in the preceding year had been dismantled and no one cared. I was determined to prove I was whole again, strong as ever. I found myself working harder than before I got sick, in a system that was no longer the one I thought I'd helped create. I felt betrayed, as much by Starzl as by the complicity of others. I told myself few of them had any choice but to go along with the new order. I began to consider what my choices might be.

# Rocket Ship to the Stars

Starzl rarely talked to me about my personal life. If something bothered me, I talked to Shun. One Monday in the late spring of 1983 he and Starzl found me in the ICU and asked if I wanted to have coffee. Dr. Starzl had a thing for the dime coffee machine in the cafeteria and he never had any change. Shun jingled the coins in his pocket and smiled at me. He hated the coffee in the machine.

Starzl and I drank ours with cream and sugar and Shun lit a cigarette. I thought Shun looked nervous.

"Shun tells me you're tired," he said.

I looked at Shun but he turned his head and blew a cloud toward the window. I'd told him late one night how tired I really was. I wished we could slow down. Maybe if I could do just four livers a week.

"Well, that's not it, really; I just—"

"These are epic times," Starzl said. "And you're right in the middle of it. The epicenter."

I said I understood that. "It's just that—"

"You're riding a rocket ship to the stars, you know. The sky's the limit. Shit, the limit's beyond the sky. It's . . ."

"Stratosphere," Shun said.

"What?" Starzl said.

"It's stratosphere."

"Yes," Starzl said and took a drink." It's stratospheric. Your career is just getting started and already it's reached the stratosphere."

I didn't know what to say.

"Well, shit, you know what I mean. I don't have to tell you how vital these times are."

Shun blew a perfect smoke ring. Starzl glanced at him, then leaned over toward me and lowered his voice.

"Look, I know you've got this new girlfriend. I know what that's like. For you."

"It's not like I want to cut back; I just—"

"But people like you, men like us need to be given special consideration. Hell, your last wife didn't do that and you were able to show her what's important."

"It's more complicated than that," I said.

"Sure it is, or it seems that way, but really, isn't that what it's about? We've got careers. They have to understand that. And you? You're riding a rocket ship to the stars. You're on a trajectory that will take you beyond anything they can understand. You know that, right?"

"There are two things going on here," I said. "This isn't about wanting to stop doing the cases. It's more about the future. I just want—"

"You want to make sure she doesn't get pissed off and throw you out. I understand. But see, that's what I'm talking about. She has to have perspective. She needs to know she's on the same trajectory with you. She can come along, as long as she under-stands what it takes."

"Look," I said. Shun dropped the butt on the floor and twisted it slowly with his heel. "I did five livers, half a dozen kidneys, and God knows how many other take-backs last week. All I'm saying is I can't keep—"

"There's ups and downs. You know that. Some weeks are busy, some not so much. It evens out."

"Too many weeks like that," Shun said.

"That's it," I said. "That's it exactly."

"So, you want to be on a schedule?" Starzl asked me.

"No, I know that's not possible. I . . ."

"You want to put a limit on how many cases you do in a week or a month or something like that?"

"No, that doesn't seem practical. I think if we had more time to train our own surgeons. We keep training people who leave and start a program someplace else and, well, I think we need to train our own sometime."

Starzl looked at Shun. "What do you say, Shun? Are you strong?"

Shun grunted.

"Can you train more?"

"I train everyone."

"Come on, now, Shun. This is serious. Bud wants to know if you can train more surgeons. To share some of this burden."

"We can all do it," I said. "We already have."

"What do you say, Shun? Are you in or out?"

"In," he said.

"Well then, that's it." He stood and knocked back the rest of his dime coffee. "What time is it, Shun?"

I told him it was a little after nine. He reminded Shun they had a meeting at nine and suddenly he was gone and Shun walked slowly after him. I picked up Shun's butt and dropped it into my cup and muddy coffee splattered my white coat.

# A Roustabout from Wyoming

Larry Heinz was a roustabout from Wyoming. Two months before I met him, he fell off an oil rig and cracked his liver.

"Out there on the Outlaw Trail," his wife said.

Mr. Heinz and I stared at her.

"You know. Route Twenty." She shifted in her seat and rearranged the manila envelope on her lap. "Wasn't wearing no safety belt."

"That's not why I fell," Mr. Heinz said. "How many times I have to say so?"

Mrs. Heinz looked down at the thick envelope. She tucked a strand of hair behind her ear, straightened her glasses with both hands, and looked up at me. "Well, this here's from Dr. Borgas." She handed me the envelope. "He said you'd find what you need in there."

Mrs. Heinz watched me set the envelope on the counter.

"Dr. Borgas sent us your records last week," I said. "I looked them over this morning."

Mr. Heinz was still looking at his wife. She scratched her arm.

"Do you have any pain?"

He shifted his weight and leaned on his other elbow.

"Larry?" his wife nudged him.

"Oh." He looked at me and smiled. "You mean me."

He told me he didn't have any pain or nausea or trouble eating or any other problem.

"Healthy as an ox," he said, slapping his belly.

I asked him to lie on the exam table. I started to open his shirt. It was one of those western ones with metal snaps and a yoke. I hesitated and he laughed, grabbed hold of each side, and popped it open in a single pull.

His stomach was flat and soft. I could barely feel the edge of his liver under his rib cage when he took a deep breath. I asked if it hurt.

"Not a bit," he said and smiled. He still looked tanned. He had a small cut on his chin. One of his front teeth was chipped. He looked a decade younger than his stated fifty-nine years.

He had a large packet of X-ray films. I took it to the workroom and sorted them by date. He'd had CT scans on three different occasions. The latest one was less than a week old. I took two of the films back to the exam room and put them up on the small viewer box.

"Have you seen these before?" I asked.

Mr. and Mrs. Heinz both shook their heads and looked at each other.

I showed them the older film first. It was taken on the day of his injury. I pointed to the lighter gray areas of the liver and said that was the normal part and that the darker area that looked kind of like a huge star taking up the center of his liver was a blood clot that had formed in the cracks in his liver caused by the fall. Mr. Heinz rose and stood next to me while his wife stayed in her chair and craned her neck.

"And that's bad," she said.

"Well, it could have been," I said. "But here he stands, alive and well."

I pointed to the more recent scan. The big star was about a third the size it was in the first scan.

"So that's good," she said.

I told them that at this point, if anything bad was going to happen, it would have. "Weeks ago," I said.

"Dr. Carey—he's the specialist back home—he was worried it could bleed," Mr. Heinz said, and looked at his wife.

She nodded, then turned to face me. "We kind of live way out in the sticks, so he'll likely be dead before help arrives."

I didn't think they needed to worry anymore. I told them about some recent literature from major trauma centers that supported the safety of not doing anything.

"So now what?" Mrs. Heinz stood up, then leaned over to get her husband's coat. "We just go home as if everything is back to normal and all?"

I suggested Mr. Heinz might want to send the younger guys up the poles from here on out. "I'm with you on that," he said.

The nurse told them to sit in the waiting room while she finished up some paperwork. I went to the workroom to dictate my note. Dr. Starzl was looking at Mr. Heinz's X-rays. I stuck the two I'd taken back on the view box.

"The man from Wyoming," I said.

He took a closer look at the original scan while I answered a page from the ICU.

He grabbed the same two films I'd used and headed for the door. "What room is he in?" he said.

I held my hand over the phone and told him I'd already seen Mr. Heinz and his wife. They were out in the waiting room. Stable hematoma. More than two months out. I said they could go home.

"Are you crazy?" He was already out the door, headed for the waiting room.

It took me a few minutes to work out the trouble in the ICU and by the time I found Dr. Starzl, he was coming out of an exam room. Inside, sitting on a chair, Mrs. Heinz was crying. Her husband sat gripping the armrests, staring straight ahead.

"He'll die if we don't operate," Dr. Starzl said. He opened the door to the next exam room. "Have Jenny set it up for next Monday."

"Even though he doesn't have any symptoms?" I said.

He paused in the doorway.

"And it's shrunk to less than half its original size?"

"Shitfuckgoddamn. This isn't a debating society." He stomped his foot and jerked his hand in that gesture, like throwing sidearm, I'd come to resent.

I went back to the exam room. Mrs. Heinz had stopped crying and she looked up when I came in the room. She had her hand on her husband's arm. Mr. Heinz stared at his shoes.

I said I was sorry. I wanted to explain that we had a difference of opinion, but I knew that didn't matter. I said someone would come set things up for next Monday. I asked if I could do anything else, explain anything. Mrs. Heinz shook her head.

That was Thursday and around midnight we had a liver transplant that would last until noon on Friday.

It didn't go well. Dr. Starzl nagged us incessantly. Help me, don't hinder me. Suck, goddammit. Don't dissect with the sucker. Goddammit I can't see. Fuck, doesn't anyone here care about life? Jesusfuckingchrist don't do that. I don't want anyone here who doesn't care about life. If you don't care about life get out right now. Now suck, goddammit. Here, hold this. Not so hard. Jesus. Come on now. You hear me? I said come on. Work, goddammit. Shitfuckgoddamn.

I hadn't slept for several days. Sometimes I'd have hallucinations, especially when things got quiet. Just little things, for split seconds. I might think I saw something flit across the operative field, or think someone had tapped me on the shoulder. Little things like that. I didn't think it was dangerous to anyone.

At one point during the transplant that night, I reached down and reset a retractor that Dr. Hong was holding, then replaced Dr. Lobretti's hands to expose better the area where Dr. Starzl was trying to sew a blood vessel. I should have known better. He accused me of trying to kill the patient. Didn't I know that he was the surgeon and that I was there to help him, not do the surgery? Did I hate the patient or want her to die for some other reason? He took the retractor out of Dr. Hong's hand, reset it, and then told Hong to take it again and pull to the sky, goddammit. He rapped Dr. Lobretti's knuckles with his needle holder, said let go, and Lobretti's hands jerked out of the field.

"Please! Someone help me!" Lobretti reached back in and exposed the field again.

With my left hand I held the free end of a suture that Dr. Starzl was using to reconnect a blood vessel on the new liver to the patient. I suddenly imagined my right fist landing with all of my strength against the side of Dr. Starzl's head. I had an overwhelming urge to kill him. Right then. Hit him so hard in the temple he'd fly against the wall and fall dead to the floor.

This was how I knew it was time for another attitude adjustment break.

That afternoon, my wife and I drove for three hours through rain and sleet to my dad's cabin in the woods on Rocky Fork Lake, in southern Ohio. The only sources of heat were a small metal fireplace, an electric heater with one of two coils that worked, and an electric blanket.

For dinner on Saturday, we decided to treat ourselves to a restaurant in town. On the way home we stopped at a Dairy Queen and got two banana splits. I said I couldn't remember ever eating one before. We ate them in the car.

We both spent that night in the bathroom. At first we took turns vomiting directly into the toilet, but then things started coming out the other end as well. I found an empty five-gallon paint bucket in the shed and we used that for vomiting, the toilet for the rest.

By noon the next day, we were both still too sick to travel. It was Sunday and I was due in the OR the next morning to help with Mr. Heinz's surgery. I drove five miles to a pay phone and made a collect call to Shun. I said I was sick as hell. Nearly dead. Banana splits. Food poisoning. I asked him to let Starzl know I wouldn't be there. He said he would. I believed him.

We left for Pittsburgh Monday afternoon and when we walked in the door at home that evening, the phone was ringing. It was the charge nurse in the OR.

"Where the fuck have you been?" she said. We were old friends. "Starzl's been calling for you all day."

Something had gone wrong with Mr. Heinz's surgery and they were going to have to transplant him. "Get here soon as you can," she said.

Dr. Starzl scrubbed out when I arrived. I think he'd been working nonstop all day.

"Oh, thank the gods," he said. "Do what you can. Münci's on his way back with a perfect liver. I'll rest in the meantime, if that's all right with you."

John was doing the anesthesia. When I stepped up to the field, the ventilator fired and a wall of blood washed over the side of the drapes and soaked my gown. I looked at John. He stared back while connecting a bag of blood to the transfusion machine. I raised an eyebrow; he shook his head.

"Not good," he said.

"When is the donor team due back?" I said.

No one spoke.

"OK, then. How long ago did they leave?"

Someone said two, maybe three hours.

That meant we had to keep Larry Heinz alive for at least three, maybe four more hours if we expected to get him off the table.

We kept working and after an hour or so we got the bleeding slowed down and I packed the cavity full of sponges. I told Hector and Eduardo to take a break, but then Starzl walked in.

"How's it going?" he said. He stood on a lift and leaned out over the abdomen and asked me to take out the packs so he could take a look. Blood was oozing everywhere and we had no clots, but I couldn't see any vessels that I could stitch either.

"You'll need to do more work there," he said. Hands clasped

behind his back, he pointed with his chin. I didn't see what he saw.

"We're plugging away," I said.

Someone stuck her head in the room about then and told Starzl he had a phone call. He took it at the desk just outside the room and I could hear his voice getting louder and louder. He was pissed off about something.

He came back in the room and stepped up on the platform again.

"Everything all right?" I said.

"What?"

"That phone call. Sounded serious."

"Come on now," he said. "Pay attention to what you're doing."

Then he left. Hector took a break while Eduardo and I stood guard, hoping pressure from the packed sponges would slow the blood, but then Hector was back and swearing.

"Fuck, shit, goddamn," he said, mocking the boss. "The donor arrested."

"Oh, Jesus," I said. "We're fucked."

"He told Münci to go get it anyway." Hector stepped back from the table and did that thing like a dice throw that Starzl does. "Münci, either you bring back that liver or I'll rip yours out and use it instead!" he imitated.

"Body's probably already in the morgue," I said under my breath.

An hour later, Münci called in on the plane radio. He yelled above the roar of jet engines. He wanted me to know that he had gotten the liver, that he thought it might be OK. I knew what that meant. Münci is nothing if not optimistic. That word, *might*, meant he had serious doubts.

Münci burst through the doors carrying his cooler with a couple of visiting Japanese surgeons in tow. He told the visiting surgeons to get scrubbed and asked our nurse to get him a gown and a set of gloves and start breaking the ice.

"Münci," I said. He paused to look at me, his doe eyes stained pink. "It's OK." Eduardo and I were sewing up the muscle layers. Anesthesia was already gone, their machine pushed back into the corner and silent.

"What happened?" he said.

"Couldn't get him back," I said. "Must have been the fourth or fifth time and nothing worked anymore. I called it half an hour ago."

I thought he might cry.

"I'm sorry," I said. "We tried."

Münci stayed around and helped us close and wrap the body in the white plastic and lift it onto a gurney. He walked with me to the lounge and I got my white coat and turned to leave.

"Where you going?" he said.

"Waiting room," I said. "To tell his wife."

# Saving Ducks

Dad and I are sitting on the patio and out past the rusty pool shed I've just seen something fly out of the hollow sugar maple on the back line.

"I think I saw a wood duck," I say.

"A what?" he says.

"Oh my God, do they still come?"

"Who?"

"The wood ducks, the ones that I hatched from eggs, what? Forty-five, fifty years ago?"

"I don't know," he says and hunches forward in his chair. I wonder if he's having trouble breathing.

"How many generations would that be?" I say. "Twenty-five? Fifteen?"

"What the Sam Hill are you talking about?"


In the spring of 1964, the first after Mom died, old man Thompson ran over a nest in his back lot. He brought the eggs to our house, said he thought he'd seen a duck fly out but wasn't sure what kind. "Wood duck, I suppose," he said. Mr. Thompson's daughter worked in the operating rooms at the hospital and I guess word of my own duck disaster had spread pretty quickly once Dad told the staff.

My duck disaster involved two dozen white Pekin ducks and a red dachshund named Vino. I walked onto the scene after an afternoon spent building a new hideout in the woods back of the reservoir. We'd been having so much fun.

One of Dad's nurses had given him a commercial-grade egg incubator. He brought it home and asked me if I wanted to hatch some chicken eggs.

"What about ducks?" I said. I didn't have good memories of chickens.

A week later some woman who smelled of ammonia and cigarette smoke brought over two dozen duck eggs. She said to keep the temperature at about 98 degrees and they'd hatch in four weeks. "Turn them every few days," she said. "After a week or so, you can candle them and throw out the empties."

Fifteen hatched but only twelve made it to the second day. I didn't have a good place to keep them, just a large wicker laundry basket with some towels spread out. I built a duck run with one-by-one lumber and chicken wire and attached it to the old doghouse, which was actually a hog house we used as a doghouse until my pointer dog, Major, got run over by a civil defense truck.

The ducks had a nice house where they could stay warm and dry and a long run that snaked around the backyard and led back to the starting point. I left the bottom open so they could run on the grass. I used some of my tent stakes and baling wire to anchor the run.

The ducks grew faster than I'd thought they would and by

the time Vino got to them, they'd lost their yellow down and were all fully feathered and white. I weighed the biggest one and he was about five pounds. I didn't name them; Dad said it was bad luck because we were just going to sell them eventually. Or eat them. He said we might want to keep a few for ourselves to eat.

I never understood how no one saw it. When I walked into the yard, Vino was lying there chewing on the leg of a duck. My clever duck run was destroyed, tipped over and scattered in pieces. Some of the ducks must have tried to get away. I found three of them in the neighbor's yard, not together but kind of laid out like he'd chased one and killed it, then ran after the other before it got far.

I ran screaming at Vino and he got up running with a duck in his mouth. I chased him as far as I could, but he dropped the duck and I couldn't keep up. I fell to my knees by the duck and started crying. I'd never seen such a terrible slaughter before. I mean, I'd seen Grandpa and Dad pull the heads off chickens and that was scary, but somehow that was different. This made me sad and angry at the same time. I thought about getting my rifle and going after Vino but I knew I couldn't do that.

I got the lawn cart and began loading the carcasses in it, and one of them tried to bite me. I dropped it and wanted to run away but I stopped and looked around and saw another one was lying on its side moving its wing back and forth. I wondered if we could save them, if Dad could help me sew them up, or fix whatever was wrong with those still alive.

I called Dad at his office. I had a hard time explaining.

"I don't understand. Vino bit you?" he said.

No. Vino got the ducks.

"How'd that happen? Did you let them out to play or something dumb like that?"

No.

I felt like screaming into the phone.

I said that it was Vino, that he must have knocked the run over, that ducks were scattered all over the place.

"Well, herd them into the doghouse and get some stakes and pound them in and use some of that baling wire and some pliers to tie the run down to the ground so it won't happen again."

You don't understand, Dad. They're all fucking dead. All but one or two and those two aren't looking so good and right now I need you to see if you can save them. Those two survivors.

That's not what I said. That's what I wanted him to understand but by then I was having trouble putting together words that made sense. He said he'd try to get home early, that I should take the live ones inside, down to the basement.

"Don't let them eat or drink anything till I get there," he said.

Only one was still alive when he got home. The one who died never looked good from what I could tell, and when I went back down to check on them just before supper he wasn't moving, so I took him out and put him in the cart with the others. I wondered if I should dig a hole and bury them. I figured Dad would know.

Dad found me in the basement talking to the last survivor. He looked it over, spreading the feathers to find the wounds.

"See these?" he said. He was showing me the upper part of the thigh. "These are puncture wounds from the canines, those long sharp teeth."

The fangs, I said.

Dad turned the duck over and found it was sitting in a pile of its own poop and there on its chest was a big tear in the flesh that seemed to go all the way to where his wing attached. Dad said the wing was probably OK but he was worried about the lungs.

"Could be this lung is collapsed," he said. He leaned down and put his ear close and listened. "Don't hear any air leaking though, so that's good."

Dad got the army ammunition box he used as a first aid kit and when he rubbed the torn skin with gauze soaked in iodine the duck went crazy. Dad yelled at me to hold him still and kept scrubbing till he got out all the dirt and leaves. He didn't use the numbing medicine like he always did on my cuts.

Won't it hurt him? I said.

He said that the stitching would likely hurt less than the cuts already did. The trouble, he said, was that some of the skin was missing and so all he could do was tack it back together here and there and then keep it clean; it would probably heal in no time.

"Ducks are pretty tough birds," he said.

I didn't know that.

The duck lived for almost a month. I was going to give him a name since he was the only one left, but I didn't. I cleaned him with soap and water and dried him with a towel every morning and when I came home from school. For a while, I thought he was being careful to poop on just one side of the box, but then it seemed like he stopped getting up to walk around, and every time I went to see him he was lying in a pool of brown, awful-smelling poop, so I started cleaning him more and more. He didn't eat much. Thelma gave me a big syringe and showed me how to use the blender to turn his feed into a kind of milk shake that I could squirt into his bill, but most of the time it just seemed to go in one side and spill out the other. He started to stink after about a week and I knew that was a bad sign. Sometimes I thought it might be better if he just died, but that made me mad at myself and then I'd try harder to make him better.

When I came home and found him dead I ran upstairs and told Thelma and she said she already knew. I yelled at her for not trying something.

You could have tried something, I said, but I knew that was dumb. She couldn't have saved him. No one could have.

Four of the wood ducks survived. I kept them in the doghouse and never let them into the run unless I was around to watch. I didn't know when they might be able to fly, but when their real feathers came in, I started taking them out one at a time to let them try. I'd get on my knees in the grass and gently toss a duck in the air. Flapping like crazy, it would sink into the grass and fall over on its head. About the time I thought they were never going to fly, one took off and landed on a low branch of the hollow sugar maple. I couldn't believe it. I sat in the grass for a while and thought it might fly back to me but it didn't. I got out the next one and it did the same thing and then all four were on branches in that tree. I waited till dark. I thought I saw one of them fly out over the field and then come back and land up higher, but it could have been a blackbird.

I don't remember how long they hung around that first summer, but once they were gone I figured that was the last I'd see of them. Two were male and two female and I imagined them going off somewhere to live together. Maybe in the creek out back of the reservoir or on an island in Lake Erie or one of those crystal clear lakes up in Michigan. The next spring, though, I saw a wood duck in the tree and before long I counted four of them coming and going. They or their offspring came back every year after that. I'd see them in the spring when I came back from college, in the summer when I worked for my dad and TJ, when I visited from Utah or Pittsburgh, and during all of those summers when we took our kids to see their grandfather.

And now I'm certain that was a wood duck I just saw fly back, probably from the reservoir.

"They're still coming back," I say. "How can that be?"

"You talking about that dead maple?" Dad says. "Been meaning to cut that thing down. You suppose you could help me?"

# Encore

I didn't know it would be my last transplant operation. I'd received chemotherapy and radiation for lymphoma more than a year before, and I told the boss I thought I was ready, but I didn't expect the call to come so soon.

I called my dad in Florida that evening. He said they'd gone out to the pizza place with the pipe organ that came up out of the floor.

"You remember that place," he said.

I'd never been there.

"Sure you have. Last spring, you and the kids."

I let it go.

"You busy?" he said.

I told him we were packing and he asked what for and I told him again about Panama.

"Panama?" he said. "Why are you going there?"

I took a deep breath and said we were going to the San Blas Islands.

"They call it Kuna Yala now," I said. "Remember when you and Mom were there?"

Of course he did.

"Well, next month it'll be fifty years after your visit."

I'd grown up watching home movies of my parents talking to

Kuna and posing for pictures. My mother had brought back a bunch of molas.

"Your mother was a sucker for molas," Dad said. "She bought so many molas, the pilot was afraid the runway wasn't long enough."

One shot in the film shows their tiny plane with its tail hanging out over the water and a couple of Kuna holding it back.

"They had to get some boys to stand in the water and hold on to the tail while the pilot revved the engine."

I laughed. I loved that story.

"Your mother made a quilt out of them. You remember that?"

"I thought she made place settings," I said.

"That may be, but she also made a quilt. You kids wrapped up in it when you were sick on the couch."

"Wow. I don't remember that," I said. "Wonder what happened to it."

"Hells bells, you threw up on it," he said.

"I did?"

"Well, one of you did. She put it in the wash and it fell apart. She was so upset."

I heard the call-waiting signal about then and looked at the ID. "Hey, Dad? I've got another call coming in. It's the hospital."

"I thought they fired you."


The call was from the operating room. They had a donor liver coming and the other surgeons were already operating; could I do a transplant?

"Ready to get back in the saddle, old man?" It was Hector, one of my partners. He was scrubbed in, doing one of the transplants, and a nurse held the phone to his ear.

"Sure," I said. I felt a surge of panic wash over me. "I can do that."

When I got to the hospital, I couldn't remember my locker combination. I found Oscar at the OR front desk.

"Hey, Dr. Shaw. You still work here?" Oscar said it without looking up.

"I don't know. Do you?"

"Not really."

"Just a pretty face, then?"

"More like sex symbol."

He stood and stuck out his hand. I grabbed it and pulled him toward me over the counter and stuck my face next to his. When I stood back he was smiling and I could tell he was a little embarrassed.

"How come you're here on a Saturday night, anyway?" I said. "Big shot like you ought to get more respect."

"Filling in's all. Lots out sick with this flu."

He handed me a slip of paper with the lock combination. "Good to have you back, Doc."

"Thanks," I said and turned to leave.

"Come on now," Oscar said. I looked back and he held his palms up like he was expecting me to say more.

"Like riding a bike," I said.

"You got that."

*   *   *

I'd never met the patient before and now she was in a coma. I could have gone up to the ICU and talked to her family to tell them I would be doing the surgery. I could have asked them if they had any questions and maybe reassured them. But I didn't. I'm superstitious. I think a lot of surgeons are, at least those who do liver transplants. Years earlier, I'd met and talked to Ellen Hutchinson and her husband right before her liver transplant. I told them she would do just fine, that I'd take good care of her and not let anything bad happen, and she died during the operation, right when it seemed everything was going along just fine. For lack of a more comforting idea, I decided then that meeting the patient right before surgery was bad luck.

This patient came from a hospital in New Mexico. None of her doctors back home had thought she was sick enough to need a transplant. We didn't know what had caused her liver to fail. It could have been some sort of hepatitis. Now she was in a coma with low blood pressure and failing kidneys.

They wheeled her into the operating room on a cart. Her eyes were open and she lay on her back. I helped them lift her onto the table. I leaned over and looked into her eyes, but she didn't look back.

"Is she responsive?" I said. I squeezed her hand.

"The ICU said she withdraws," someone said.

I pinched the nail on her index finger and she closed her eyes and pulled her hand away.

"Sorry," I said.

They put a small pad under the patient's head and covered

her with a blanket. I went over to the door and looked at the thermostat and turned it up to eighty-five degrees because blood doesn't clot well when it's cold. I went back to my stool.

The anesthesiologist gave the patient some drugs to put her to sleep, paralyzed her, put a tube through her mouth and into her trachea, and connected her to the ventilator.

"OK?" the nurse said.

Julie was the anesthesiologist. She nodded and the nurse stripped the blanket and gown off and the patient was naked. I got up and helped get her into position. We put foam pads on her arms and legs and wrapped her legs with sheet wadding so they'd stay together. We put a pillow under her thighs so that her knees bent slightly.

Julie had a resident and a respiratory technician with her. They worked on threading catheters into veins and arteries and hanging bags of fluid and cross-checking identification numbers on bags of blood.

"Put on a gown and gloves for that," she told the resident when, bare-handed, he started to wash the area over the artery in the woman's left wrist. He sighed and went looking for a gown pack as Julie pulled gloves on over the sleeves of her gown and looked at me. "Fucking residents," she said.

"Ill-mannered barbarians," I said. "Every one."


When I dictated my operative note the next morning, I said that we freed up the liver in the usual manner. I remember thinking I'd done well. Danielle was my assistant that night. She said she thought everything had gone amazingly well. Like I'd never been away, she said. But not everything went as planned.

*Within two to three minutes of releasing the portal venous clamp, the patient's heart stopped,* I dictated.


"What's the potassium?" I said.

Julie was under the drapes by then. I felt her head pushing against my elbow.

"Sorry," I said. I realized my elbow had been banging against the endotracheal tube. I twisted a bit more to my left so that I could continue pumping on the chest without dislodging the tube that ran down into the patient's trachea. "Is it OK?"

She stood up, took a quick look at the oxygen saturation monitor, then looked at me. "It's all good," she said.

From my side of the table, it had gone from all good to utterly horrible when our patient's heart stopped.

"So what is it?" I asked again.

She stared back.

"The potassium?"

"Oh. Yeah, jeez. Let's see." She grabbed a clipboard, the one on which I'd just seen the lab tech attach a slip of yellow paper. "Uh, four point one."

"Well, that's not it," I said. "Maybe it's air."

Danielle and I pumped on the chest for what should have been plenty long enough to get any air out of the heart so that it would start pumping again. "How long's it been?" I said. I was getting tired.

"Oh, about forty minutes, maybe less." Julie looked at her clipboard, then at her watch. "Thirty-eight minutes, to be exact."

"Your turn," I said to the medical student.

After a few minutes I looked at the blood pressure. It wasn't as good as when I'd been pumping.

"You need to get higher," I said. "Get the man a couple of platforms."

Danielle took over pumping while the student backed away and someone put a couple of lifts next to the table.

"Climb up there and get right over the top of her chest so your hands are going directly downward," I said. "You want to trap the heart between the sternum and her spine and really mash it."

The student came down on the chest with all his weight. A gush of dark blood rolled out over the top of the liver.

"Whoa!" Danielle said. "Maybe a little *too* hard."

I looked at Danielle. She grabbed the sucker and shoved it into the wound. It made a lot of noise and the tubing filled all the way to the canister mounted on the wall.

"Where's all that fucking blood coming from?" I said.

I turned my headlight into the belly. The blackness of the river welling up over the incisional brim and onto the drapes told me we had a leak in a vein—a big leak in a big vein.

I grabbed a handful of white cotton pads off the nurse's stand and shoved them in above the top of the liver, hoping to put pressure on the leaks. The liver, previously pink, looked like a bloated eggplant.

We waited. I could hear the student grunting and when I looked at Danielle she was grinning.

"That's good," I said and nudged the student with my elbow. "Pressure's solid."

He took a deep breath.

I pushed Danielle's sucker tip in a little deeper and a noisy stream of burgundy ripped up through the tubing again.

"OK, we've got to fix that," I said.

Danielle took over doing the heart massage and I shoved my hand into the space above the liver and pulled out the packs. They were dripping black-red. I sucked out the pool inside and caught sight of the suture line; each time Danielle came down on the sternum new blood boiled up through the torn strands of vena cava.

I asked for the suture and looked at Danielle, who glanced at the student and nodded. I checked the blood pressure monitor. "Looks good," I said.

"OK?" she said.

"Yup."

Danielle paused and I dove into the slot between the dome of the liver and the diaphragm. The student was slow with the sucker.

"Can you see?" I said.

He pulled out the sucker and craned his neck and blood covered the line.

"Neither can I. You need to suck down there or I can't fix this mess."

Danielle and I worked silently together. She paused, the student sucked out the blood, I placed a stitch, pulled up on the line, and Danielle pumped again. We did that maybe a dozen times and at one point the student used two suckers and Danielle looked at me and, short of breath, shouted, "You da man!" to the kid and like that, we fixed it. When we were done the student took over for Danielle. After about another twenty minutes the black blood was coming up again and I had to fix it all over again.

"The cava's so ratty. Can't do that a third time," I said. "You suppose we've got some tamponade?" I spoke to no one in particular. Danielle had been pumping for a while and I saw dark blood welling up again.

"Fuck. Cava's going to fall apart for good." I grabbed the electrocautery. "Time to open it up."

The medical student stopped sucking and looked at me.

"The pericardium," I said. "She might have fluid inside, acting like a vise on the heart." I grabbed the sucker and cleared a small amount of dark blood. "Probably not, but it's worth a try." I handed him the sucker. "And you saw how easily the cava is tearing, right?"

He nodded.

"So, we need to stop this jumping up and down on her chest and get a hand on her heart."

"Like internal massage," he said.

"Suck down there, OK? I'm going to need to see."

Alternating between burning my way with the cautery tip and probing with my finger, I wormed my way under the lower edge of the sternum, working to time my probings with Danielle's compressions. I concentrated on not going too deep and making a hole in the heart. When my finger popped through it bumped into the wall of the heart.

"Fuck. No fluid here," I said.

I inserted two fingers into the hole, then my whole hand. I

wrapped my fingers around the heart. It was flaccid—not even a twitch.

"What's he got now?"

Julie looked at me and I indicated the EKG monitor with my chin.

"Still flat?" I said.

"Yep. We had some runs of V-fib there, but they didn't last long enough to bother."

I started squeezing the heart, but it was too big. I couldn't get my hand all the way around it. I shoved my other hand in and compressed the whole heart between my palms. I looked up at the clock. We'd been without a spontaneous heartbeat for nearly an hour.

I thought the patient could survive the night. Most of the time, she appeared to have a good blood pressure during the external massage. We had one episode during which she appeared to have a heartbeat, but most of the time she didn't. We performed both external and internal cardiac defibrillation three or four times.

"Hold on a minute." Julie grabbed the side of the cardiac monitor and turned it slightly to get a better view against the glare of the big lights.

I was sitting on a stool by then, my gloved hands in my lap, my neck bent, my eyes closed. Danielle was squeezing the heart again. I'd just checked the time and realized we were now past ninety minutes with no signs the heart was coming back. I was about to call it quits.

"We've got something!" Julie said.

I stood up and took my place opposite Danielle. I grabbed her arm. "Hold on," I said.

We stared at the monitor. The flat line that had marked our time for so long had become a slowly rolling series of smooth, round waves. In seconds, they grew taller and began crowding closer together until, like a TV show cliché, a normal electrical complex sprang like a spark from the end of a fuse.

"Fucking A," Julie said.

"Feel anything?" I said.

Danielle felt the heart.

"Un-fucking-believable," she said. "We've got a beat." She pulled her hand out and we stood watching as a normal cardiac rhythm spread across the full width of the screen and marched on undaunted.

"Pressure?" I said.

"It's all good," she said.

At that moment I loved no one in the world more than I loved Julie.

By three o'clock in the morning, we were still in the operating room and the patient appeared to be improving. She was making urine, and the liver started making bile. We got the bleeding stopped, closed her up, and took her to the ICU.

The evening after the transplant, I felt compelled to write an e-mail to my friend Dirk, in Utah. We were leaving for Panama in a few days and needed to finalize our plans. Dirk's e-mail goes through the censors at LDS Hospital but I'd learned how to avoid their rejection of my messages. When they bounced a message containing "boob," as in "that phukhead is a total boob," I discovered that "booob" slipped by unmolested. I ended up writing a tedious recounting of my last night in the OR. In the last paragraph I wrote . . . *but despite all that, she's awake and responding to commands. The gods have spoken. Welcome back motherphuquer.*

We left for Panama two days later. We were gone nearly two weeks, island-hopping in sea kayaks, guests of the Kuna—sleeping on their beaches, eating their food, singing their songs, marking their stories, honoring their boundaries.

When I came home, I never made a decision to stop doing liver transplants, but I turned down the offers to do another until my partners stopped asking. I told myself it was better for me, that going without sleep so much could let the lymphoma back in, that I'd done enough and now the younger surgeons could do as well, maybe better, without me. At least for a while. I didn't yet understand that would be the end of it.

# PART THREE

# Remission

# Superhero

I didn't want to be there. I stood in the grand ballroom at the Holiday Inn and all I saw was a blur of people wearing name badges with red numbers. I'd operated on many of those with numbers. I'd taken out their broken livers and put in new ones. They were back in Omaha, gathering on a searing Saturday in July to celebrate being saved. It was a milestone anniversary of our first liver transplant and I couldn't stand being there.

I hadn't been to a reunion for at least ten years. I decided to go again because of the anniversary thing. Chris had gone with me to many of the past reunions and some of the older patients and their families had grown fond of her. I knew they'd be looking for her and that I'd have to explain that after twenty-five years, we'd recently divorced. Before we left the house, I told Rebecca that I was dreading it more than cancer. How melodramatic, she said.

Standing there in the ballroom, I felt Rebecca watching me. She took my hand and squeezed it; I took a deep breath and waited. Any second someone would recognize me and call out my name and start that dance, the dance of dread.

I bet you don't remember me, do you?

You remember Johnny. He was number twenty-nine!

You don't know my name, do you?

I ain't seen you for long as I can remember. Stay there and I'll go get Hannah. Boy'll she be surprised.

We had our first patient reunion in 1986. We invited patients and their families, nurses and doctors, administrators and social workers, anyone who'd been involved in the program. The local media outlets came and stories of jaundiced babies washed white by shiny new livers led off the ten o'clock news.

For ten years, I scheduled our family vacation so that I'd never miss a reunion. I don't know what changed, but ten or twelve years on, I couldn't bear those reunions. That's when I came to a mature relationship with my anxiety, said fuck it, and stopped going.

As cover, I made sure our family reunions in Ohio always conflicted with the patient reunions. Our head nurse complained. She thought I had a responsibility to be there.

"I've got no control over when my family decides to have these things," I said.

"Right," she said.

"Besides, all that attention gives me the creeps."

She said I was being ridiculous.

I shook my head.

"They all just want to see you," she said. "So they can thank you."

"Yeah, well, Hannah won't leave me alone."

"She loves you, Bud," she said. "You saved her life. You're her superhero."

I used to call the reunion Hannah's party. Hannah was two years old when her liver quit working. The same thing had happened to her brother and he'd died. By the time Hannah came up from Arkansas she was in a coma. Her brain was swelling quickly and without a new liver, she would die.

Hannah got her new liver. She had a long recovery but went home well and happy. She was three years old when she came to her first reunion, and then and every year that followed, she wouldn't leave me alone. She'd track me down and follow me wherever I went.

She loved to give me her chattel: balloons, cookies, washable horse tattoos, pieces of fried chicken, copies of the program, finger paintings, a forkful of potato salad, stuff she found or made. One time I went to the restroom and suddenly there she was leaning against the urinal. She had one of those balloon dogs a clown had made for her, holding it up to me as an offering.

I might have saved Hannah, but when I watched her laughing and covered in finger paint I also remembered Heather. Heather was eight years old when I gave her a new liver. Two years later, a surgeon in Oregon called me at three a.m. He had Heather in the operating room and wasn't sure what to do.

"We're a small county hospital," he said. "We're not used to this kind of surgery."

He told me about her black intestines, her swollen liver, and her failing blood pressure.

"Sounds bad," I said.

"Yeah. I'm afraid she's not going to last much longer," he said.

"Maybe if you took out her intestines her pressure would come up."

He said he could try that. "I'll talk to her mother, see if that's what she wants."

Heather never made it out of the operating room.

A day or so after Heather died, I received a letter from her. Inside was a picture of her sitting on a rock at the top of a mountain in Oregon. Her mom had taken her on a hike to celebrate the second anniversary of Heather's liver transplant. "I'll miss seeing you at the reunion this year," she wrote. "See you next year."

I got little comfort knowing that Heather was a thousand miles away when she died. My being with Kelly, another patient, had done her no good.

I'd replaced Kelly's liver when she was three years old and a year later, her mother called us to say Kelly was short of breath and feverish. Sounded like the flu, but with transplant patients, I go for the worst-case options first and ask questions later. I said they should take her to the emergency room and get her started on Bactrim because it could be pneumonia, the kind caused by a parasite. A week later, Kelly was on a ventilator in a hospital in Illinois and the doctors still weren't treating her the way I thought they should, so I grabbed a colleague and we flew to Illinois and fussed over Kelly in the ICU. It was all too late. We had a conference with her doctors and I nodded a lot and said they were doing the best they could. Kelly was dead by the time we landed back home that evening.

Once patients went home, I couldn't do much to protect them. I couldn't prevent hepatitis or cancer from coming back, and too often it did. I couldn't keep them from drinking again and I couldn't make sure they took their medicines. Their future would always be filled with so much uncertainty.

I looked for Hannah at the anniversary reunion, but Laurie said she couldn't come, that she had a new job and couldn't get the time off. I felt both relief and disappointment.

Jane Morgan was there. Jane never misses a reunion. To celebrate the anniversary she made a quilt that included an example of each year's T-shirt. I used to have a say on the design of each year's T-shirt, and we came up with some brilliant slogans back then: "We De-Liver," "Liver Longer in Liver City," "Liver City Round-Up." My favorite was "Let's Paté"; it took me four years to get that one approved.

Jane hugged me and I looked at the 21 on her badge. Everyone knows his or her number. Jane was our twenty-first person to get a liver transplant.

"This must be twenty-four years for you," I said.

"Twenty-four and a half," she said.

She handed me a marking pen and asked me to sign her quilt. She said she wanted to get all the original gang to sign it.

Rebecca and I stayed to watch the pictures. We used to get all the patients in one shot, and I was the photographer. Now, a man with a microphone sorted them into groups, calling out the year of transplant and type of organ. We saw moms and dads holding babies, old men with canes or walkers, boys with belts sagging nearly to their knees, girls in tiny shorts and too much makeup, young men and women looking courageous, empty-nesters laughing at a father with a screaming two-year-old. Somehow in all the chaos, everyone got properly immortalized.

I let them down, I thought, watching the madness. People like Jane. I'd been such a coward. All those years, I'd made this about me, about how these people made me panic with all their talk of miracles and saving their lives.

I watched Jane climb onto the stage. She stooped to help a boy onto the step, then stood and waved at her husband whistling in the crowd.

"This is what it's really about," I said.

A year later, I got a wedding invitation from Hannah. I thought I should go—to represent our transplant program and all—but I put the invitation away and forgot about it.

Hannah sent me a picture of her in a wedding gown. She looked healthy and strong. Bulletproof, I thought. She and her husband celebrated their first anniversary by being baptized in their home church. A few months later, Hannah died.

At home, I told Rebecca about Hannah; she asked me how I felt.

"It's got to be hard," she said.

"No," I said. "I'm OK."

She took a sip of coffee, set the cup down, and watched me.

"Seriously," I said. "It's not like I could have done anything."

That evening, I found Hannah's obituary on a funeral home website. She loved working as the manager of a local equestrian farm. She excelled at showing horses and had recently won reserve champion with a Percheron mare, Hannah's Storm, at a national event. According to her father's wishes, Hannah was buried in her favorite Razorback hoodie, her worn-out chaps, and her old cowboy hat. "Now she can ride easy," her dad said.

Years have passed. I haven't been to another reunion, not since the anniversary. I couldn't sleep recently and sat in the predawn darkness looking through my folder of patient photos for pictures of Hannah. I've been trying to write about her and the reunions, trying to understand why, for so long, I've avoided them and why Hannah became an excuse. I keep looking at her photos and cards, thinking I've missed something, something important to this story.

Sometimes I have this flashback of Hannah. I'm in front of the TV cameras doing an interview with the governor about our wonderful transplant program and there she is, tugging on my hand, holding up a finger painting that looks like mustard on a napkin. I ignore her, not by doing anything overtly hostile, but in not turning away from the reporters to reach out and accept her gift, in not lifting her up and telling her how much her gift meant to me.

# February 25, 2014

## Stinson Beach, California

I'm trying to get by without the drugs again. This isn't the first time. I woke up this morning shaky and anxious, like something horrible was about to happen. I slept a good eight hours last night, maybe more. No dreams, no nightmares. So why the fuck do I feel like all is lost, that I can't possibly get through the day? Why does thinking about coffee scare the crap out of me? I love coffee. I fall asleep thinking about how good that first cup will taste in the morning. I can almost smell it. Now here it is morning and the thought of drinking something fills me with fear. I'm afraid of what might happen. I tell myself I know what this is, that it happens all the time, that I just need to stay calm, make a pot of coffee, drink a cup, maybe two, and it will go away. And sometimes it goes away, and if it doesn't I can take a Xanax.

In the *Times* this morning there's a story about that ship that was hijacked in 2009, the one they made that Tom Hanks movie about. Seems they found just two guys dead inside. Navy SEALs, or rather, ex-SEALs. Big, muscled guys, I guess. Guys who worked out a lot. They found one of them faceup on the floor, still staring up at the ceiling, a syringe in his hand. His left hand, the story says, and I wonder if he was right-handed. It says they found brown heroin in the cabin. The guards were on shore leave in the Seychelles and had been partying all night; someone said they got back to their ship at six o'clock in the morning. An ex-wife and a bunch of friends and neighbors all said they were good guys. No way they did drugs, they said. I know the Seychelles islands are somewhere in the Indian Ocean, but I Google them and zoom in on the harbor, trying to imagine where the ship was docked when they died.

I can't see myself partying all night in the Seychelles—or anywhere, for that matter. I look up from my computer and see that the waves are getting closer to this cottage we've rented on Stinson Beach, in California, and I remember we have friends coming over at noon for lunch and I'm terrified of walking down the beach to that restaurant, our favorite place here, and I just don't know why. Something inside me does this to me and I wish I could make it stop.

# Surviving Botulism

One night during my first year in medical school I came down with botulism. This wasn't my first serious disease.

In my first month in medical school, I was convinced I had rectal cancer. It started with a low-grade pain in my anus. It felt like I had something up there, something bigger than a golf ball but smaller than a grapefruit. More like a plum.

The pain was gone every morning. Thank God that's over, I would say, looking in the mirror. I always overslept, so I tried to drink my coffee while I shaved. The problem was the shaving cream around my lips. I thought about using a straw, but I could never remember to steal some from the cafeteria.

We sat in a lecture hall every day from eight till noon. By ten o'clock I'd have finished my third cup of coffee and would be nodding off. We had breaks every couple of hours and I think I slept through most of those as well. By lunchtime my butt was aching and I knew it was only a matter of time before I got the bad news. It's cancer, they'd say. You'll need surgery.

I knew what that meant, from the summers I'd worked with TJ. The first time I helped him do *that* kind of surgery I couldn't figure out why he'd placed the man's feet in stirrups, like they do when they take out a woman's uterus. He started in the belly and got the colon all freed up as low as he could, then he went around to the foot of the bed, sat down on a rolling stool, and disappeared between the man's legs. He told me to pull up on some metal retractors while he carved out the guy's anus. I was kind of off to the side, leaning against the patient's leg, and couldn't see much for a long time, but then he told me to go back inside the belly.

"Reach down there and grab the rectum," he said.

It was dark in there. TJ had moved the lights to shine on his part of the operation. I reached up to move a light.

"Leave it," he said. "You don't need to see. Just grab it and pull it out."

I felt around down there. My hand bumped into a metal clamp and I grabbed it and lifted it out. It reminded me of the time my dad took me fishing in the ocean and I caught a big jellyfish.

When I looked back down where the rectum had been, I could see the floor. Suddenly, Dr. TJ's head came into view. He looked up at me and waved.

I used to have nightmares in which I was awake and having my rectum excised. Sometimes they ended when I fell through that hole.

I eventually went to the student health clinic and told some guy who looked like Eli Wallach about my rectal pain. I would have told him it was cancer but I didn't want to bias his opinion. He said he needed to do a sigmoidoscopy. I'd seen that done to a person as well. It involves a shiny steel tube the size of a broom handle. That's when I thought I'd been overthinking the plum-in-the-ass thing, and by the time he was done and told me he hadn't seen anything up there I wasn't relieved; I was ashamed to have been so afraid.

The botulism was a little different because I didn't have a lot of time to worry about having disfiguring surgery. With botulism, death is more or less immediate.

That first year, I lived with a black dog named Shadow in a sooty gray apartment in Cleveland's Little Italy. It was owned by widowed twin sisters. On the last Friday of every month, they invited me upstairs and served me warm anisette and biscotti and asked for their check. For months on end they were my only human contacts outside of school.

"You pay us now," said the one, standing. They wore matching black dresses and I could never tell them apart.

I bought my groceries from neighborhood shops. The best deals were smelt for twenty-one or twenty-two cents a pound and chicken backs for eighteen.

One night I got home after all the shops were closed. It was November and dark and I still didn't have a car. What I did have was a box of Cap'n Crunch in the cupboard and two beers in the fridge. The beers were behind a plate of leftover chicken backs. I couldn't remember how long they'd been in there.

The chicken didn't taste right but I didn't mind much. After it was all gone, I sat there watching TV. Shadow was on the bed licking himself but I was too tired to yell at him. Then my stomach started making a lot of noise.

One of the lectures that day had been on the pharmacology of bacterial toxins, nasty substances that can make you really sick or, like in the case of tetanus, usher you through a slow and horrible death, like what happened to Mick Hidy when we were in third grade and he stepped on a rusty nail. Or like botulism, which, as I recalled hearing in the lecture, you can get from eating bad food.

I tried to concentrate on the TV. Hawkeye Pierce was saying something to Hot Lips Houlihan that must have been pretty funny, but part of my face was going numb. Botulism can do that, I thought. They said something about paralysis that starts with the muscles in the eyes and face.

I looked toward the bedroom. Shadow had two tails. Standing in the bathroom, I looked to see if my face was drooping. One eye looked like it was more open than the other but it always looks that way. This seemed worse than usual. I grinned to see if both sides moved together and noticed a twitching along my left eyebrow. "These could be signs of botulism," I said to my melting face. I thought about what came next: difficulty breathing, stumbling gait, generalized flaccid paralysis. If I didn't get help in time, I'd stop breathing and suffocate while wide awake.

I took Shadow for a walk. The leaves were down and blown into little piles along the curbs and against the storm grates. I remember wishing I'd worn a hat.

The cold made my face and hands numb and weak and even though I'd convinced myself I was crazy, after ten minutes back in the warm apartment without regaining sensation, I knew I had to do something.

The nurse on the phone at the emergency room told me they were really busy.

"You don't arrive in an ambulance, you going to wait a long time. You might come down later when this mess clears."

Like in a half hour? I asked.

"Like in the morning, honey."

Things got worse. Pretty soon my hands were cramping and I was having more and more trouble breathing. I started to feel dizzy and thought I might pass out.

Shadow got up and went to the bedroom. I heard him turning round and round on the bed until he stopped and let out a low groan. I turned off the TV and sat staring at my bookshelf and saw my copy of Davis's *Microbiology*.

The cover made a cracking noise as I opened it; the pages were clean and crisp and stuck to each other. As I read the chapter on *Clostridium botulinum*, my breathing slowed, the tingling in my face and fingers faded, and I felt that growing sense of euphoria that comes with great new discoveries, like the scientific facts my textbook contained that made it clear why I couldn't possibly have botulism.

They'd probably mentioned all this in class, I thought.

So just like that, I was cured. At least that time it didn't take having Eli Wallach shove a metal tube up my ass.

# June 1, 2002

## Torrey, Utah

Why had I never felt it before? It was big. Not yet a golf ball, but close. And hard. I felt a surge of fear and a pounding in my throat and leaned the top of my head into the tile wall of the shower.

I let the warm water flow down my back, down between my legs, watching it vortex down the drain. I had to be mistaken. This thing would go away, this big, hard lump in my groin.

Joe, my fourteen-year-old son, and I checked out of the motel and drove an hour east through Capitol Reef National Park to Hanksville, then followed the directions they gave us south over dirt roads to the wedding. The groom's father was a cancer surgeon and one of my best friends, but I didn't tell him about the lump. After the wedding, Joe and I drove south into the Waterpocket Fold and spent two days camping and hiking. Twice I decided against walking routes that the guidebook said might be a little treacherous. Such newfound caution was foreign to me and I found myself apologizing to Joe.

"I don't know why, but something tells me we shouldn't go up that way," I said, and we stayed down low and missed what must have been a spectacular view. At night I couldn't keep myself from running my fingers over the lump, testing its size. Sometimes I thought I could feel parts of it extending like fingers along the vein into the deeper tissue of my thigh, but then I'd feel it again and decide I was wrong, that it was just a round lump and really not that big. I wanted to fall asleep listening to Joe's deep breathing. I wanted his calm and serenity, his optimism, a future.

We carried our own water and didn't come close to running out, but early on the third day I said we'd better get back to the car. The road back took us through red-rock canyons past heartbreaking beauty, and I wanted to stop and photograph it all but I didn't. I was desperate to get back. We stopped only once and I took a picture of Joe sitting on the lip of a swirl of striped sandstone that looked like a giant conch shell split open by time.

# March 2005

## A Worst-Case Scenario

I've always been neurotic. That's probably what fuels my obsession with worst-case scenarios. That can be a good obsession to have if you are a doctor, as long as you can sort out the truly irrational possibilities from the ones culled by knowledge and experience and common sense.

I think being a little neurotic, worrying about the worst-case scenario, made me a better doctor. If your belly hurts, it's most likely something you ate. Maybe you're constipated. You know you don't get enough fiber, right? Or it could be the flu. If you come see me, I'm going to want to make sure you don't have appendicitis or diverticulitis or stones in your gallbladder or a kidney infection or pneumonia. Those might be pretty common causes of your pain. My worst-case scenario alarm, though, won't let me not worry that you have cancer of some sort, or infection in the pancreas, or some sort of poisoning or a twisted bowel. Those aren't that common, but they're far from rare, and you don't want me throwing those possibilities out before we get some more information, get some blood tests and maybe an X-ray or two. But you don't want me thinking seriously that you are the next person in the United States to present with an exotic viral infection that starts with abdominal pain and leads to fever and a rash and diarrhea, then bleeding from every orifice before your lungs fail and your heart stops and everyone around us who tried to save you is now infected and going to die as well.

Crazy? I don't think so. Nothing's impossible. I just need to keep the crazy stuff in check until the ranks of the not-so-crazy things begin to thin because all the tests for them come up short.

Then? Then you might want someone obsessed about worst-case scenarios. Like me.

The first time I felt the new lump in my axilla, I ignored it. It wasn't easy, but I did it, ignored that voice shouting a warning that this was my cancer come back to kill me, to lead me down a path of more chemotherapy, a bone marrow transplant, a series of uncontrollable infections, and a horrible end with a tube in my throat and a ventilator hissing.

Like Emily Dickinson's fly.

I heard a Ventilator hiss—when I died—

So I ignored the lump just long enough to get on a plane that evening and fly to Utah, where I was scheduled to give a lecture and receive an award for being pretty good at what I'd done in my career. On the flight to Salt Lake City, I reached up under my arm about every ten minutes. It's not so big, I'd tell myself. It's soft, and I can roll it around. That's not like a cancer. If it were cancer it'd be hard and immobile, fixed like a burr with its hooks.

I thought of that night in 2002, three years earlier, camping with Joe. That time I was able to corral the wild beast that is my diagnostic mind. I did what I could to keep from rubbing the lump as I lay staring into the impossibly star-filled abyss of our camp sky.

When I deplaned in Salt Lake City, I had a lot less evidence than in 2002 that something serious was going on. Even so, by the time I collected my skis and rented a forest green Subaru, it was past midnight and snowing relentlessly and I was full of dread and absolutely certain this would be my last time to travel anywhere.

I took the 600 South exit off I-80 and headed east through a downtown full of slush. The streets were empty, but three blocks past State Street, I had to slam on the brakes to wait for a man in a navy hoodie to stagger across. At 700 East, I had the light to turn left but some bozo in a Bronco ran the light, found me already in the intersection, spun out, and slammed into the curb opposite Trolley Square. In my rearview mirror I saw him back up and drive off in the direction from whence he'd come. Stopped at South Temple, past midnight, snow falling in clumps like dryer lint, I had to wait for a man wearing loafers, a wool topcoat, and fuzzy carmuffs out for a walk with his dachshund. Its legs out of sight, the dog seemed to float by.

I was convinced these weren't coincidences. These little events were all signs. My cancer was back and I was going to die soon, but not before unimaginable suffering. I arrived at my friend's house, slipped in through the garage, and fell asleep on the guest bed without undressing, counting my breaths, pacing my heartbeats against the pulse of history and the looming end of time.

I attended the honorary dinner, accepted an award, smiled

and acted thrilled. I skied the next day with old friends and on the mountain I found myself soaring through knee-deep powder, aware I had never skied better, had never felt so free, so immortal. I took that as a sign as well and when I called home that evening I told Chris I wanted to come home. She and Natalie, our daughter, were supposed to fly out to stay with me the rest of the week but I didn't know how I could stand it.

"I'm terrified," I said and explained the lump.

She said it was just as well. Natalie had a fever. Some bad flu going around, she said.

I hung up and called my oncologist. He was in Shanghai but would be going through Omaha in two days on his way to Lyon and could see me. I had a biopsy a few days later and he called me from Lyon to tell me it was nothing.

"Just fatty infiltration," he said. "Sometimes we see this."

I hung up the phone and waited to see what would happen next.

I ought to cry, I thought. I should be blubbering with joy. Why couldn't I cry? Why couldn't I feel anything?

Then Natalie came in and showed me the blood and I forgot about being so close to death. I couldn't prevail over something that never existed, but now my daughter had coughed up something tangible, a blood clot no bigger than a fingernail.

# Exploding Cigarettes

Before my mother got cancer, my dad was already an antismoking fiend. He made speeches at churches and schools. He brought a movie to my school that showed these machines smoking cigarettes so they could collect the smoke and make this brown goo out of it. One guy in the movie pulled a mouse out of a cage and used an eyedropper to feed the goo to the mouse, which then wiggled like crazy and died, and just like that the guy dropped it into a glass beaker. Another movie showed some lab doctors in white coats painting the goo on the backs of mice and not long after the mice had these gross warty lumps all over them. Dad said it was cancer and that they'd all be dead soon.

At home, Dad complained all the time about my mom's smoking. He said she smoked more than two packs a day, sometimes three, and that if she didn't stop she was going to die. My mom didn't ever see any of the movies, and she didn't come to the pancake breakfast for the Cancer Society, even the year after she already had cancer and I stayed up all night to help set it up. Grandma Kinnear had been smoking all her life and she was an old lady. Aunt Fran, Mom's sister, smoked all her life and she was sixteen years older and still didn't have cancer. "But then there's your father," my dad would say when she argued about it. He had smoked all his life and he died of cancer right before Mom and Dad's wedding.

Mr. Stapleton owned a neighborhood grocery on North Street. We used money we earned mowing and raking and doing other stuff to buy pop and candy and, after I saw those movies about smoking and dead mice, cigarette loads.

Cigarette loads were these little white things about the size of a kernel of long grain rice. It was just some gunpowder wrapped in thin paper; Mr. Stapleton sold them in red-and-white metal tins with ten loads each for a quarter.

I figured that if I could stuff a few cigarettes in every pack with loads, my mom would never know which ones might explode or when, and so she'd have to quit out of sheer terror. I could always find a pack lying around and would stuff five or six cigarettes. I used a toothpick to shove them into the tobacco at the end so that she'd never be sure when one might go off.

That was the year I got a pup tent for my birthday in February. On some Saturdays, Dad and I would put it up in the living room and my friends Jim and Teddy would come over and we'd sleep in it. It was up most of the summer in the yard just across the driveway from the kitchen window, and that's where we'd run to right after we loaded up a pack of Mom's cigarettes. We'd wait there despite the heat, playing fish or war with broken decks of cards, and when we heard the bang, we'd run in through the garage laughing. Mom was so mad. One time it caused a small burn hole in the gray dress she wore to work at the hospital in Chillicothe. Another time she said it burned her eye and I said maybe now she'd quit. But she never did.

# July 1963

## Alone in the Back

I wish I could remember the fireworks on the Fourth that year. Eight days later my mother was dead. She was in the hospital for a few days before that, so I doubt she saw them or anything else that week. Not that fireworks were all that great in Washington Court House, but still. I wish I had a crystal clear memory of her sitting in a lawn chair in Eyman Park, swatting mosquitoes and waiting for the final moment when they light the Fiery American Flag and we all stand and sing "America the Beautiful."

Right after lunch on the day they would take her away, I went to the city pool. I sat with Patty and her friends and I worked my way up to lying beside her with my arm across the small of her back. I had a pretty good boner by then and when she rolled over and told me it was time to get cooled off and headed for the water, I couldn't roll over. Not yet. Back then we wore those super-tight trunks and I knew everyone would see it and that would be the worst thing that could happen all summer. Maybe all year.

Patty stood on the edge of the deep end and turned back to see what was keeping me. "Come on!" she yelled, commanding me with a wave.

"In a minute!" I yelled back. I was still on my stomach and things weren't getting any better. I rolled over and pulled the towel over me and sat up with it covering my lap. She looked pissed off. I didn't want her to be pissed off. I needed to do something, so I thought about my mom standing over that mess on the new carpet, the robin's egg blue carpet.

There was an ambulance out front when I got home. Two guys stood on the front porch with a stretcher. Their uniforms were all white and they were waiting for something to happen.

Dad came to the door and held it open and they took the stretcher inside and I knew they'd come for Mom. Dad came out and waved to the driver. He didn't look at me and I wasn't sure he knew I was there. The ambulance backed up over the curb and onto the yard. The grass clippings were thick and stuck to the tires and I wondered if Dad would notice I hadn't raked them up.

Mom's eyes were closed when they brought her out, but she wasn't sleeping. She had her fuzzy blue blanket pulled up close under her chin, like she was cold. The guys in white stopped on the porch and looked at each other and one of them said something. They tried to lift the cart up and carry it down the steps but one of the wheels caught on the concrete. I thought they were going to tip it over but Dad yelled and grabbed the side just in time. Mom tried to sit up and her hand came out from under the blanket and grabbed Dad's wrist.

They had a hard time pushing the wheels through the grass clippings. When they got to the ambulance, they lowered the stretcher to the ground. They lifted it and slid it into the back of the ambulance and cut grass fell like rain from the bottom of the cart and blew onto their white pants and shoes. I saw Mom wince from the jerks and stops, and then she pulled the blanket up over her mouth. Her lips looked blue, darker than the blanket. I could see the yellow glow of her baldness just before they closed the door. Dad headed for his car and the ambulance bumped away over the curb, my mother riding along, alone, in the back.

# Mr. Hardin

Mr. Hardin is dressed and ready to go home. He's hugged all the nurses and his grin is a miracle. I didn't think this day would come. He was so sick when he got to us, took so long to recover. He lost a lot of blood during his transplant, more than most, but now he's going home.

Well, not *home* home. He'll stay in town for a few weeks so we can check up on him, so he's close by if something doesn't look right.

He sees me with the team and calls out my name. We're making morning rounds and everyone turns as he walks down the hall toward us. He stops suddenly, holds out his arms, looks down at the floor, and then collapses like someone has cut the imaginary strings holding him up.

When I lay him out on the floor I see that he's vomited blood. Bright red blood, the fresh stuff—not old brownish blood that leaked out a while ago. Something just let loose. His belly looks tight to me and I can't feel a pulse in his neck.

The resident starts pumping on Hardin's chest and someone says they'll get the oxygen and a bag. I wiggle my finger into his skin wound. It hasn't healed that well yet, what with all the prednisone we've had to give him for rejection, so I easily split it open by running my finger up and down. I'm able to cut the blue plastic suture that holds his muscles together and his belly ruptures open with a gush of bright blood, so I shove my hand deep up under his liver where the artery comes in and I can feel a quick, soft rushing.

"Fucking artery's come apart," I say. "Someone get a cart."

The crash cart appears.

"Any of you know how to intubate?"

No one moves.

"You!" I point at someone I think might be a medical student. "Get in here and hold this for me."

He kneels down on the floor opposite me and someone hands him some gloves and he can't get them on and I can tell he's not going to get through this but then I can feel a pulse that isn't in time with the resident's pumps.

"We've got a pulse," I say. "Stop pumping."

I tell the resident to try bagging Hardin first to see if his breathing gets stronger, and it does. I've got control of the bleeding, he's got a pulse, and he's breathing again.

"Call the OR," I say. "Tell them we're coming, ready or not."


I discovered that the artery to his new liver had become infected and ruptured. I used a piece of artery from elsewhere to bypass the infected segment and hoped the antibiotics would get rid of any remaining infection in the area. Mr. Hardin made it through surgery that day and two weeks later he was ready to go again. I saw him walking in the hall the evening before he was scheduled to go out.

"Got any surprises in store for tomorrow?" I said.

"Lord, I hope not, Doc. Believe I've had enough of surprises."

I shook his hand and wished him good luck, in case I didn't see him in the morning. He stopped me as I started to walk away.

"Say, Doc, I got this here weird pain on my left side." he said. "That anything to fret about?"

"Weird how?" I said.

"Oh, I don't know what you'd call it." He lifted his shirt and pointed to a place on his rib cage below his left nipple. "Kind of a stitch, maybe. Especially when I cough."

I put my fingers on his chest and gently pressed on one rib, then the next, and suddenly the rib gave way and he let out a scream and nearly fell to the floor. I held him up and looked around for someone.

I'd broken his rib with just gentle pressure. Or it was already broken. I was never sure. We got him back to his room and gave him narcotics for the pain. I told him he'd feel much better in a day or so. Two days later, the nurse found him dead in bed. He'd had a fever and we thought the infection was in his lungs because he wasn't breathing deeply enough from the pain in his rib, but we were wrong. The bypass artery had become infected and it too ruptured. The broken rib and all those narcotics were coincidental; he'd bled to death in his sleep.

# March 2005

## Is This Normal?

My sixteen-year-old daughter coughed up blood an hour after my oncologist told me I didn't have recurrent cancer.

"Is this normal?" she said, holding a tissue with a glob of red phlegm.

I was at my desk writing an e-mail to a friend telling him I was not going to die after all, and she said she'd been coughing a lot and said that her chest hurt and that she felt really tired.

"And I can't stop shaking," she said.

The emergency room doctor showed me her X-ray and told me he was a bit surprised by how extensive it was.

"We don't usually see a lot of kids with pneumonia this bad," he said.

He seemed a bit nervous, maybe because I was the chairman of his department, but at the time I worried he wasn't telling me something. He gave us some antibiotics and said to come back if things got worse, or didn't get better. That was Saturday night. On Monday, I was in the clinic when I got a call saying she had a bad bug. It was MRSA, the nurse said, but it was sensitive to the pills she was taking. "So that's good," she said.

A few days later, another doctor called me and said he thought Natalie should go on intravenous antibiotics. I said she seemed to be doing fine.

"That may be," he said. "But the standard treatment for MRSA requires intravenous antibiotics."

We spent half the afternoon waiting for someone to insert an IV catheter. I thought I could do it myself in ten minutes if they gave me an operating room, some local anesthesia. and a fluoroscope. They set up a home nurse to come by every day and check on Natalie and we went home that evening.

I gave Natalie her morning dose of antibiotics, starting the drip at about five o'clock and running it in over an hour. She got dressed and her mom took her to school. I went home early that first afternoon and found her on the couch. The visiting nurse was there to check her IV catheter and give her another dose of antibiotics, and when she was done taking Natalie's blood pressure and pulse, she wrote it in the chart and asked her how she was feeling.

"OK, I guess," Natalie said.

I looked at her more closely and realized her lips were a little dusky and she seemed to be shivering. I asked the nurse about Natalie's temperature and she said she hadn't taken it yet.

"And the blood pressure?" I said.

She looked at her papers.

"Fifty-eight," she said.

"What do you mean?" I said.

"I could only hear it at fifty-eight," she said. "Maybe I should take it again."

This time she said it was 59.

"Is your blood pressure usually this low?" she asked Natalie.

I was on the phone with the emergency room by then and told them I was bringing my daughter in with septic shock. In the car, Natalie said she'd felt really tired all day at school.

"I could barely walk up one flight and I had to stop and rest at each landing," she said. "And I was late to some of my classes."

It took several hours and half a dozen liters of intravenous fluid to get her blood pressure above one hundred. A chest X-ray showed the pneumonia was worse. I called my favorite infectious disease consultant and the ER staff called a chest surgeon. They came up with a plan and once she looked like she wasn't going to die, they put her in the hospital.

She went to a new ward. It was billed as an intermediate care unit, for people who needed more attention than those on a regular ward, but less than on an intensive care unit. The staff was also mostly brand-new nurses. They were very cheerful.

I was surprised at how terrified I was. I couldn't leave her room and sat watching the monitors for bad signs, trying not to show my concern to Natalie. I worked very hard not to be a meddlesome physician-parent. She fell asleep watching a movie involving horses and teenage girls.

About midnight, Natalie's pulse suddenly increased above 140 and she woke up short of breath. I called the nurse and she took another set of vitals and left the room without saying anything. After ten or so minutes, I couldn't stand the fear and went out into the hall looking for the nurse. I found her in the nurse's lounge shopping online.

"Hi," I said.

She looked up and smiled.

"Was her blood pressure OK?" I said.

"Who's that?"

"Natalie," I said. "The patient in room 709."

"Let me check."

She found the chart and told me her blood pressure was 75. "I couldn't get the lower number," she said.

I explained that Natalie had come into the hospital in septic shock and that the numbers suggested she was getting sick again.

"Well, the doctors ordered antibiotics and she's getting them. What would you like me to do?" She closed the chart and leaned back in her chair and stared at me, her head tilted.

I told her to call the doctors, that Natalie needed to get fluids and get them soon and probably get lots of them.

"I'll see what I can do," she said. I thanked her and went back to room 709. Fifteen minutes went by and Natalie's heart rate rose above 160. Her breathing seemed shallower to me and she had trouble talking between breaths.

"Am I going to die?" she said.

# A Man in Control

I was a good surgeon. At my peak, I felt I had few peers in the operating room. Whether I did or not wasn't and will never be important. I think it only mattered that I believed I was the best. I think that was the only way I could get through seemingly impossible operations; it was also what made me good at teaching many other surgeons to do the same. I developed a routine that I insisted was the right one for doing a liver transplant, and I changed those steps only when I saw a need for improvement and only in a careful, systematic fashion. I was never afraid of change, but to maintain standards, I felt strongly that change almost always needed to come from a careful review of experience, not from whim.

From what I know of other surgeons, none of this is unusual. The operating room is our fiefdom, the only place in our lives where we can demand supreme command. It's where we direct the troops, fight the battle, and, in trade for such vivid autonomy, win. Mastery merely requires us to know what the hell we're doing when we step up to the table and wield such hairy power.

I've always believed I was a good transplant surgeon, in part because I worried as much about what happened after the surgery as during it—that perpetual obsession with worst-case scenarios. In transplant surgery, you learn to suspect any symptom could be a harbinger of something lethal, and you do whatever is necessary to seek and destroy the thing that wants to kill your patient. The more times a patient confirms your worst suspicions and the more often your quick action saves him or her, the smarter and more powerful you feel. Though often less dramatic, it's the same thing that goes on in the operating room. You convince yourself that you can fix anything, that you can control what happens, that your aggressive actions are justified both by the risks and by your own stunning brilliance.

It usually worked. When it didn't—when I lost someone despite my best effort—I could usually find something or someone to take the rap, at least in my personal summary of what happened, the summary that let me off the hook and allowed me to maintain that necessary sense of power and control.

These days, I don't control anything. I haven't found anything in the real world to replace my operating room. In the years since I stopped doing surgery altogether, friends and colleagues, mentors and former students continue to ask me if I miss operating and I struggle to give them an honest answer.

I miss the reward, I say; the feeling of elation—ecstasy, actually—when a long battle comes to an end and we're all bloodied but alive. That usually gets a knowing nod.

I miss the joy of doing something wonderful with my hands, I tell them. I built a wooden kayak a decade or so back and that was fun, but nothing like sewing a liver in and letting the clamps loose and watching the liver fill with blood, turn pink, and start making bile before my eyes. Everyone smiles at that. We've all had concerts in which we never missed a note.

And I miss the camaraderie, I say. Being part of a team that works together to pull off something impossible, highly skilled professionals working side by side, doing their best. Too often that causes a pause, and sometimes someone will laugh and shake her head as if to say, "If only," or "More like a clusterfuck where I work," and I begin to remember how rarely things went so well. I usually stop then. These seem to be the answers they expect to hear, so that's enough. If I go on, I lose them. But it's not the whole story, nor the most important part of it, at least for me.

In 2005, though not taking on transplants myself because I didn't want to do night call anymore, I was still working full-time as a surgeon, operating regularly and taking my stints at rounds. I was well into the ninth of twelve years I'd serve as department chairman, and I'd started working on a new software development project.

That was the year I spent much of the last week of winter in my daughter's hospital room. I was sure that if I left, something bad would happen. More often than not it did.

I was there on the first night, when the nurse didn't recognize that Natalie was going back into septic shock and needed resuscitation. I can't bear to consider what might have happened had I gone home to sleep in my own bed. I wasn't there a few days later when they gave her an exceptionally large slug of Compazine by vein, which made her unresponsive for most of the next two days. I'd gone home for a shower and new clothes and a decent lunch and felt punished for doing so. On another occasion, when I went to my office for twenty minutes to sign some overdue papers, they accidentally pulled out a critical intravenous catheter while helping her out of bed and sent her off to radiology to replace it with only her signature on the consent form.

But Natalie got better. She went home, missed her final swim meet, and got to be on TV because she was one of a few people who'd had the same kind of staphylococcus infection and, luckily, hadn't died. I was convinced my being there helped save her life; everything that happened during those weeks only worked to reinforce the notion that I could control fate as long as I was willing to be there. Always. I wasn't able to consider the impossibility of that.

Natalie graduated in May and as reward, I took her horseback riding in Quebec. Chris and I dropped her off in August at college and drove the twelve hours home in near silence, each of us struggling alone to comprehend the change. A few weeks later we cycled with friends on back roads from Budapest through Slovakia to Poland, toured Auschwitz, and fell in love with Krakow. A traditional Thanksgiving with family in Ohio, Christmas at home in Nebraska with kids home from college, and I was aware of little beyond the wonders of good health and the usual politics of a university job.

On a Sunday afternoon in January 2006, I was working at home in Omaha, writing a paper for an upcoming presentation. The Pittsburgh Steelers were in a playoff game with the Indianapolis Colts on the TV in the den and from what I could hear the Steelers were ahead. Most likely I'd be done in time to watch most of the fourth quarter from the couch, or go for a bike ride if the score wasn't close. Things were good.

With no warning, I began to feel anxious. I stopped writing and took a few deep breaths. I felt the pulse in my wrist and it was bounding and fast. I began to feel a little light-headed and when I stood up, my legs felt weak. I went into the family room and sat down on the couch. I couldn't comprehend what was happening in the game. Maybe I was having a stroke, I thought. Or I had a brain tumor. Or maybe a brain aneurysm was leaking and I'd be an organ donor before the fourth quarter even started.

I knew that was ridiculous and I wanted to laugh at myself, look in the mirror and tell myself to cut it out. Relax. I found my wife and told her I was feeling strange. "How so?" she said, but I couldn't explain. It was January but 60 degrees and sunny, so I thought I should take a bike ride. A short one, I said. It would help me relax.

After a few blocks I knew I'd made a mistake. I was obsessed with death. Everything I saw or felt filled me with dread. I hated the sun. The crazy warmth made me shiver with fear. A fit young woman jogging by in shorts and a bikini top in January was a signal that the world was about to end, if not for her and the others, at least for me. Everything I'd ever done in life was meaningless and I was the only one who knew what was coming next. I turned around and rode home, went straight to my bedroom, and crawled under the blankets. I was freezing cold and I covered my head to make it dark. I lay on my side, brought my knees up into my chest, bent my neck, and made a tent of the blanket so that I could breathe, but when I felt like I was suffocating, I stuck my head out and sucked and sucked for breath.

I had another attack a few days later. It also surprised me because it came out of nowhere. I wasn't feeling the least stressed or worried about anything. Over the next week, I had four or five more attacks. I felt cold and had weak legs and the thought of any contact with anyone or anything utterly terrified me. When it happened at work, I told my staff not to bother me, drew the blinds, and locked my door. I lay down on the couch and covered myself up with my winter coat and shivered until I fell asleep, and when I woke up I was euphoric from the realization that it had passed again.

I didn't know what to do. I wondered if I had some hormonal disturbance. It could be my thyroid, I thought, or maybe I had a weird tumor that secreted substances that made me feel this way. Several times I felt dizzy and that made me worry it was something in my brain, so I called up the head of neurology and asked him what he thought. He made me get an MRI of my brain, drew some blood, and wrote a note that said he was worried about recurrence of lymphoma in my brain. The tests were all normal and he didn't know what I should do next. I thought that maybe I should see a psychiatrist, but I didn't want to bring it up.

I eventually discovered I had an anxiety disorder. A friend told me, surprising me with a confession of his own struggles. Treatment helped bring it under control, mostly. I had breakthroughs now and then, though, and for that I took a small dose of Xanax or Ativan. Several times I had to cancel a scheduled surgery or ask a colleague to fill in for me. I never operated after taking a sedative, even the tiniest dose. I knew I wouldn't have any problems doing so, but what if I did? I wouldn't take that risk, no matter how low.

I've come to wonder if control was never more than an illusion that I created to allow me to survive. Even so, the notion that someone would die if I wasn't there to do the right thing had been a valuable, if ultimately quixotic fiber of my being for so long. That's who I was, after all.

# Trauma Call

He didn't mention trauma call. I'd have remembered that. But there I was, first day at work as a new attending surgeon, or a real surgeon as the residents say, and my chairman handed me my copy of the general surgery and trauma call schedule.

"What's this?" I said.

"Call schedule," he said. "I let you off easy this month."

I argued that I was going to be up all night a lot and shouldn't have to take trauma call, but the chairman said it was a small program and not much happened on call nights anyway.

"It's no big deal," he said. "Not like we're in Dallas or Houston, you know."

I think I felt a little ambushed.

"You never mentioned this before," I said. "You know, back when we were talking about things like this."

Sitting in my office later I decided it probably wasn't going to be a big deal, at least not for a while, not until we got busy, and so far we hadn't even put a patient on the transplant waiting list.

I told my partner later that day. He seemed happy to have something to do for a change in between the long wait for donors.

"Trauma can be fun," he said.

I stared back at him, surprised. I hated trauma.

"I'm sure once we get busy he'll let us off call," he said.

He didn't. More than once I found myself coming off thirty or more hours without sleep and relieved to be heading home when my pager called me to the emergency room. The busier we got with transplants, the more resentment I felt. I brought it up with the chairman again, and I found myself admitting that trauma call was pretty quiet, that it had yet to interfere with our transplant work.

My toddler son got his hand caught in the jamb of our front door and Chris called me at work to say the tip of his middle finger was hanging on by a thread of tissue. I was finishing a liver transplant at the time and told her to take him to the ER and I'd meet her there as soon as I was able.

A few minutes later, the on-call organ procurement agent called me to say they had a donor in our hospital. They wanted to go to the OR as soon as possible, he said. The family had given consent and the child was very unstable, was I far enough along to break away and do the donor surgery?

I looked at the clock and told him I'd be another hour, maybe less, and he said that'd work out fine. I had the nurse in our room call to get the recipient for the baby's liver headed our way, to update the family of the recipient for the next liver, and to see if my wife and son had arrived in the ER yet.

And then things got complicated.

The surgery chief resident came to the OR to present the case of an eighty-three-year-old man with diverticulitis and a perforated colon.

"What do you want to do?" I said without looking up. I was in the middle of suturing the bile ducts together.

"Well, Dr. Shaw, I reckon he needed to be in the OR yesterday," he said.

I explained my situation and told him to see if he could find one of the general surgeons to help him.

"I kind of knew that, so I already tried and no one has answered my pages."

I asked him to call the chairman.

"I'm afraid he's out of town, sir."

I said to go ahead and get the man in a room and I'd figure something out.

I eventually found myself running back and forth between three operating rooms, making sure I was present in each one for all of the critical moments. As I was finishing up the liver transplant, they wheeled my son into a fourth room. I watched them transfer him to the operating table. He had an intravenous catheter stuck in the back of his left hand and he wasn't moving. I walked over to see if I could talk to him, tell him he'd be OK, but he was already out cold. I put my hand on his forehead and looked at his tiny eyelids and felt the warmth of his skin.

"Shouldn't be a problem," someone said and I turned to see the plastic surgeon putting on his magnifying loupes. "Kids do surprisingly well with this injury. He'll be good as new."

I went next door to make sure the old man with the ruptured colon was stable and talked to the chief resident about how he planned to start. I told him I'd be next door doing the donor and to call me once he got in and took a look around and knew what he wanted to do.

"OK," he said. "But you should know I just got another call from the ER. They have a fourteen-year-old they say has a ruptured appendix."

"Give it to the fucking pediatric surgeons," I said.

"Can't," he said. "The kid's over the age limit."

"For what?" I said.

"We've got a rule says fourteen and over goes to general surgery," he said.

I told him to let me know when they got him in a room, the kid with the ruptured appendix. In the meantime I needed to get the donor operation going.

On the table in the donor operating room was a small boy with blond hair and pink skin and black eyes and a broken skull. He was my son's age. All I could see lying there was my son and a wave of panic came over me, filling me with a sense of doom. I felt my thighs go weak and I began to shiver and then I rushed out to the sink, pulled down my mask, and heaved and coughed and held on to the metal sides to keep from falling.

# Don't You Want to Save Her?

Not everyone who needs a liver transplant will get one even though more people support organ donation than ever. That means we have to choose who gets a new liver and who doesn't, and to be fair, to best serve the public good, we need to be picky. From the earliest days of organ transplantation, almost every transplant center set up committees composed of surgeons and medical specialists, nurses, psychologists, psychiatrists, social workers, and others to review the individual candidates and to decide who could go on their waiting lists for donor organs. Congress passed a law in 1984 that led to the creation of the first national standards for the use of donor organs for transplantation, including the development of national policies related to both organ distribution and patient selection. The national rules allow individual transplant centers some discretion in their selection of transplant patients.

Some causes of liver failure can come back and destroy the transplanted liver. An alcoholic can go back to drinking, a hepatitis virus still in the recipient's body can infect the new liver, and a cancer in the liver can appear to be cured but come back anyway. All of these recurrent ailments can destroy the new liver, sometimes in a matter of months, other times not for years.

The first goal of an organ transplant is to cure the disease that destroyed the native organ. We know that's not always possible; but even without providing a cure, we can still give patients years of high-quality life after an organ transplant. Of course, that leads to a thorny question: How long is long enough? Is a year of survival for Mrs. A enough when hundreds of other people on the waiting list are likely to live for decades after getting a new organ? What about two years?

We're also not that good at predicting how long someone might survive before the original disease comes back to get them. In general, if the selection committee members conclude that the patient has a better than fifty-fifty chance of living at least two years without the original disease destroying their new liver, all else being equal, they'll usually put them on the waiting list.

These decisions are often far from straightforward. No one on the selection committees around the country wants to turn down anyone with a shred of hope. It was always the toughest part of the job for me.

Carmine Williams had a liver transplant for hepatitis when she was twelve years old. When the pathologist looked at her old liver, he also found a tumor. It was a liver cell cancer and small enough that we told Carmine's father it was unlikely to come back. Two years later, her hepatitis was back but not the cancer, and a year after that, we decided to give her a second liver. She did well at first, but after only five months, her liver was failing again and a CT scan showed what looked like cancer in the newly transplanted liver.

I met with Carmine and her father, Hoyt, in the clinic on a Thursday afternoon. I told them the committee had turned

Carmine down for a third liver transplant. Carmine just stared at me.

"But that other doctor said that without a new liver, she wouldn't make it to Christmas!" Hoyt yelled. "What do you have to say about that?"

"That's hard to know," I said. "But even if we give her another liver now, either the cancer or the virus will just come back and—"

"Do you know that for sure?" he asked. "Hell, you're the doctor who told us last week you're not even sure about the cancer. You don't even know where it's from, do you? I mean, for all you know, it came with that liver you put in her. None of the tests have proved a damned thing."

"We're pretty sure—"

"Pretty sure? You're pretty sure? You give my daughter a death sentence based on pretty sure?"

I waited to reply until he scooted back in his chair. Carmine leaned back and shook the hair out of her face.

"I know this is hard—"

"Oh no you don't. Don't you dare say that. You have no idea."

"Mr. Williams," I said. "Let's go over what we know for sure."

"I can't believe this," Hoyt said. "She's come so far. Look at her. Look at this sweet, sweet little girl. Don't you want to save her?"

Fifteen-year-old Carmine sat leaning forward, elbows on her knees, staring at the floor.

"Goddamn it, she *deserves* another chance," he said.

Hoyt said that if I wouldn't help her, he'd find someone who would. I said I could refer them to another transplant center. I knew of one or two who might accept her. If that's what they wanted, I'd get on the phone that afternoon.

"But if Carmine were my daughter, I wouldn't want her to go through that," I said.

"We shouldn't have to go somewhere else," he said. "You're supposed to be the best here. I know. I did my research."

"I'm sorry," I said.

Hoyt turned away, wrapped his arms around Carmine's shoulders, and she buried her head in his chest. I left them there like that. I didn't know what else to say.

The day after I gave Carmine the bad news, I was sitting at a desk in the ICU looking over a patient's lab results when I got a page. It was the hospital attorney, asking if I'd ever heard of Clinton Walker.

"Clint Walker is an attorney representing a patient of yours named Williams," he said. "He called the hospital administrator, suggested we turn on channel twelve at noon."

The anchor on channel 12 called it breaking news. "We take you now live to the university hospital where an attorney says doctors have condemned a Beaver Falls girl to certain death."

Walker and a reporter from channel 12 stood outside in the sun. A gust of wind caused the reporter to grab her hair. Clint squinted. The large metal letters spelling out UNIVERSITY HOSPITAL formed a perfect arc over the two of them.

Mr. Walker told the reporter from channel 12 that the doctors at the medical center had condemned Carmine Williams to death. For no good reason, he said, the doctors were denying Carmine her right to survive.

The reporter wondered why.

"So do we," said Walker. "Mr. Williams and his daughter don't understand. How can a reputable medical center like ours have such merciless doctors?" he mused.

In the afternoon, the hospital attorney, practice group attorney, and my boss and I met with Walker to prepare for a press conference. The conversation started out civil enough, and then it grew heated. I sat there wondering how so much energy could be expended with so little understanding. Neither side seemed to have all the facts.

"You know," I said, "we offered to refer them to another center."

Mr. Walker stopped in mid-rant, his tongue suddenly stilled. The hospital attorney rubbed his face with both hands.

"When was that?" asked the practice group attorney.


All four TV networks, three radio stations, and two newspapers sent reporters to the press conference. Clinton Walker announced that he had reached an agreement with the hospital lawyers.

"I've worked out an acceptable compromise," he announced. "The doctors have finally agreed to refer Carmine Williams to another center."

The channel 12 reporter, her hair immobilized, asked the department chairman, my boss, if that was true. He looked at me then and I said that I had already called a transplant surgeon in another state and that surgeon had agreed to see Miss Williams.

"What are the chances she'll get a new liver?" she asked.

"I can't say for sure," I said. "Probably good."

"Why can't we do it here?" she asked.

"She doesn't meet our criteria for another transplant," the chairman said.

"But won't she die?" someone asked.

"Yes, she will," I said. "Either way."

"How so?" someone asked.

"We won't discuss those details," my boss said.

"Are you angry they're going elsewhere?"

"Of course not," he said. "We're relieved."

Hoyt and Carmine left a few days later. A couple of months went by before one of our nurses copied me an e-mail from Hoyt. Carmine had gotten her new liver; she was doing great, the best she'd been in years. He hoped we'd learned our lesson so others wouldn't have to go through what she had.

Carmine lived less than a year. I didn't find out what got her in the end. I worried that Hoyt was surprised they had so little time after all that pain and suffering, but maybe he thought it was worth it. I know I didn't feel the least bit vindicated. I kept thinking we needed to figure out a way to do better by people like Carmine and Hoyt.

# Vigil for Malcolm

I didn't really know much about lawyers until I was in medical school. During our first year, we played touch football against the law school boys. Most games were friendly enough, but one of our guys, Eric, liked to pick fights. "Hey, sleazeball, what do you call five hundred lawyers on the bottom of the sea?" he yelled at their quarterback halfway through our first game. "A good start!" He thought he was funny. After that he didn't have to use the joke anymore, just the punch line. "A good start, mother-fucker," he'd say whenever he felt the urge, like after tagging someone hard enough to knock him down. Some of my class-mates thought it was funny. I guess I just didn't understand the rule that if you're a doctor, you automatically believe that all law-yers are weasels.

In the early days of my career as a transplant surgeon, I worked with various support agencies to help patients obtain fi-nancial approval for liver transplantation. Sometimes I had to testify in federal court on behalf of patients when state Medicaid agencies refused to pay for liver transplantation. We always won a reversal, and in the process I got to know several remarkable attorneys who, like me, worked those cases for no pay.

I also had to defend myself against claims of malpractice half a dozen times over thirty years. Those suits always named my partners, the hospital, the nurses, and dozens of people peripheral to the events. In most instances, the cases were dropped during the discovery process. On the few occasions that my colleagues and I went to court, the juries decided in our favor. I learned that lawyers really could be your best friend, at least when going up against other lawyers. Those experiences also taught me something disturbing about the power of truth. The trick in court is to bring truth into plain view, and I saw that lawyers could be experts at both ensuring and thwarting that.

Eventually I learned that even when the truth becomes evident, it isn't always relevant. I also learned that once in a while, I talk too much.

I spent many nights in the ICU during the week Malcolm Dial lay there recovering from a stroke after a liver transplant.

Some nights, Malcolm's sister, Mary, and her boyfriend kept me company. I'm not sure when I first met Mary. She wasn't in the waiting room after surgery when I told Malcolm's mother everything had gone well. Mary arrived days later, after I told Malcolm's mother her son had suffered a major stroke and might not survive. I rarely, if ever, saw Mary during the day. She and the boyfriend would roll in after a night out, smelling of booze and cigarettes, and flop into Malcolm's ICU room. Most nights, the boyfriend would be asleep before his head hit the red vinyl of the recliner.

Mary was less predictable. One night she seemed excited, certain Malcolm was getting better. She roamed the room, watching the monitor, checking out the urine bag.

"His blood pressure's better, right?" she said. "And look at all that urine. That's a really good sign, right?"

On another night she pulled a chair up beside Malcolm's bed and sat holding his hand.

"We grew up together in Mississippi, you know. Oxford. William Faulkner and all that," she said. "We spent all summer fishing and gigging frogs up on Toby Tubby Creek."

She stood up and leaned down and kissed her brother on the forehead.

"We were so close back then," she said. "Seems so impossible we've come to this."

She collapsed into the chair then and sat sobbing and shivering. The nurse handed me a box of tissues and I set it on the side table next to her.

"Things were never the same after Daddy died," she said. She told me they'd had a falling out over Daddy's will, something about a last-minute change that pissed Malcolm off and that he blamed on her. I was too busy adjusting Malcolm's ventilator by that point in our conversation to listen to all the details, but I recognized the guilt of estranged siblings when one shows up late to witness the death throes of the other.

One night Mary arrived shortly after midnight. She was alone and sober and she wanted to talk. A lot. Somewhere in that rambling conversation she asked me why Malcolm had a stroke.

"I know you talked to Malcolm's doctor," she said. "But I didn't understand what he was talking about. Why don't you explain it to me? You know, in layman's terms."

Having a willing audience, I told her how surprised we all were, how rare a stroke is after a liver transplant.

"Huh," she said. "So what do you think went wrong with Malcolm?"

"I wish I knew," I said. "I've looked back over the data and everything and I can't find anything to explain why it happened."

Malcolm started coughing and I could hear fluid in the tubing from the ventilator.

"Excuse me," I said.

I got up and dumped fluid from the tubing, causing the alarm to go off temporarily.

"Is he OK?" Mary said.

"Yeah," I said. "I think he needs a little suctioning."

I put on some gloves and went to get a suction tube when the nurse came in.

"I can get that," she said.

Mary had risen from her chair and stared at her brother's face.

"Is he in pain?" she said.

"He's fine," I said. "Might be a little fluid in his trachea, but that's all."

Mary sat back in the chair and closed her eyes. The nurse finished suctioning and Malcolm's breathing settled nicely. She checked the Foley bag, where Malcolm's urine collected, and wrote the number on a piece of paper towel and left with it.

Mary sat up and the recliner snapped forward with a bang. She leaned forward with her elbows on her knees, hands together, her head down like she was about to invoke the power of the gods.

"So no clue. You have no clue what caused his stroke."

It was late at night and we were alone together, keeping watch over her brother's struggle to survive. I felt safe. I wanted to share my only theory about Malcolm's stroke.

"Well, there may be one thing," I said.

I told Mary how sometimes an air bubble inside the new liver, or in the blood vessels, can cause a stroke. "I've never seen it," I said, "but I read a report from another transplant center." Their patient had a hole inside the heart between the chambers, which they thought allowed a bubble to bypass the filter of the lungs and travel through the aorta and right into the brain, blocking the blood flow.

"A hole in the heart?  Isn't that really bad?"

"Not usually. Not if it's small," I said. "And not if you don't somehow get air in your veins."

She frowned at me and shook her head.

"A fetus has holes between the chambers of its heart that allow the blood to go the right direction. They're supposed to close after birth, but sometimes they don't close all the way. If the hole is small or not in a critical location, a person can go through life perfectly healthy and never know."

"Unless they get a liver transplant," she said. "Like the patient at that other center."

"Well, but that was a little different. They already knew about the hole."

"How?"

I explained that they had done a complete heart evaluation before the transplant and found a small hole but didn't think it would be a problem.

"Just shows you can't be sure there aren't any bubbles left," I said. I told her that the other center speculated that they could have prevented the stroke by fixing the hole before the transplant.

"So what about Malcolm," she said. "Does he have a hole?"

"Turns out he does," I said. "But we didn't know about it ahead of time."

Mary stared at me then, not with the blank look from nights at the bars, but with an intensity I suppose should have made me wary.

"So, you didn't know about that ahead of time," she said.

I said I didn't. "Like I said, it probably wasn't the cause, you know? People have strokes for lots of other reasons."

"But you checked him out for all those other things and didn't find anything to explain it, right?"

"Yes, but—"

"Well, let me ask you this. Didn't you check out Malcolm's heart ahead of time?"

I explained that Malcolm's age was below the cutoff we use to decide if a patient needs a full heart evaluation.

"Do you wish you'd done that test, what did you call it, the—?"

"Echocardiogram."

"Yeah, that. If you'd done it and saw the hole in his heart beforehand, would you have gone ahead with the liver transplant?"

"Hard to say," I said. "Maybe not."

"Maybe not? I can tell you I wish you'd found it and fixed it. From what you've said, seems like it was the only likely reason Malcolm's lying here like he is, not moving, not talking, not doing much of anything that you could call human."

I got up and got the nurse's bedside chart and tried to look at the data, but the numbers were a blur. We sat together listening to the air rushing in and out of Malcolm's lungs. I couldn't resist Mary's argument. I wish I had known about the hole. I hadn't told her that I'd been the moderator of the session where a surgeon from the other center had presented the case of a stroke at an international congress. I'd even commented in the question and answer session that I thought maybe we needed to consider routine echocardiography on everyone before liver transplant, but the presenter and others argued that was overkill. "It's such a rare event," one said, "how can you possibly justify the expense?"

"Come on. Tell the truth, Bud. If you had it to do over, you wouldn't have done his transplant without fixing that hole first, right?"

"The way I feel now? No. I wouldn't do it if I knew he had a hole, and we easily could have looked for it ahead of time."

I don't recall any other conversations with Mary. Malcolm began to recover rapidly, Mary went home, and after nearly two months, we shipped Malcolm off to a rehabilitation center near his home.

Malcolm had been living at home for about a year when I got a phone call from my former practice group's attorney. Did I recall Malcolm? Did I remember talking to his sister, Mary? Did I know that Malcolm's rehabilitation ended up costing quite a bit? Did I know Mary was a lawyer back in Mississippi whose firm made decent money suing doctors?

Plaintiff's counsel, who was someone Mary would later tell me was "the meanest dog in the south," named me in the suit.

They also went after every physician and nurse who had anything to do with Malcolm's initial evaluation, the university that employed us, and the hospital where the surgery took place. We ended up in court after years of discovery and depositions.

I flew back the night before I was required to testify and arrived at the courthouse early the next morning. Malcolm and Mary sat on a bench waiting for the doors to open. I stopped and said hello. I realized then that Malcolm and I had never really met before. One of my former colleagues had performed the original evaluation months before the transplant, and by the time I'd arrived to do the transplant, Malcolm was already asleep on the operating table, being prepared for surgery. When they sent him to rehab, he had yet to regain any clear capacity to remember. So I had to explain who I was while Mary stood looking at her shoes. Malcolm smiled and nodded. Mary began to tap her foot, then looked at her watch.

In the courtroom Malcolm seldom looked at me during my testimony. I always caught Mary looking away just in time. Or maybe I imagined it. Maybe I just wanted her to feel something—a little guilt, maybe. We'd spent time together, Mary and I. We'd become comrades, up in the middle of the night: Malcolm's sister, Malcolm's doctor.

On the witness stand, I answered a series of initial questions from the plaintiff's attorney meant to let the jury know who I was and a bit about my experience. Then he got down to business.

"So when Ms. Dial asked you whether you would have performed a liver transplant on her brother, Malcolm, had you known he had a hole in his heart, what did you say?"

"I said I probably wouldn't have."

"Probably?"

"She asked the same thing that night. . . ."

"And—"

"And I said I would not."

"You would not have done the transplant until Malcolm had the hole repaired, is that correct?"

I nodded.

"Doctor, I need a verbal answer."

"Right, that's what I said at the time, but I was wrong—"

"Objection, Your Honor—"

"—since the chances of a stroke are so small that—"

"*Objection*, Your Honor: nonresponsive, foundation."

"Sustained," the judge said.

The judge told the jury to ignore my comments. He told me to stick to the question asked. The lawyer asked me again if I said I wish I hadn't done the transplant, my attorney objected that I'd already answered that question, and the judge told the plaintiff's attorney to move on.

I don't think Malcolm got anything out of his lawsuit. The defense brought in experts who argued that the cost of doing a heart echo on every candidate wasn't justified by the very rare chance of a stroke. The jury apparently agreed and although they thought we ought to be better at figuring out who could get a stroke, they didn't think we'd broken the standard of care.

Malcolm was still alive many years after his liver transplant, and years after I left the institution. He lived near Mary on the gulf coast of Mississippi.

At a national transplant conference one year, I ran into a nurse from the transplant team who told me they were still closely involved in the management of Malcolm's transplant medications. The nurse said most of the communication with Malcolm went through Mary. Mary was the person they called when Malcolm was overdue for his lab tests, or to make changes in his medicines.

I was glad Mary cared about her brother. Malcolm's recovery was never complete, and he needed Mary's support.

But I wish I knew more. I recently wondered how Malcolm was really feeling. Was he happy? Was he glad he got a liver transplant? Could he have told me, or would I have had to ask Mary for her opinion?

One evening, I used Google to find an address that I figured was Malcolm's and put it into Google Maps. I used the street view mode to get a better look. It showed a small, pale yellow box of a house. The door of a single-car garage was open. A gray SUV sat in the driveway, its rear hatch open. I could see two paper bags sitting on the cement to the right of the rear bumper. Across the street, a neighbor was power-washing his driveway. The neighbor next door had half a dozen rotting cars and trucks parked on the crabgrass under rawboned trees.

I felt strange spying on him. I'm not sure what I was looking for. I wanted these images to be alive and fresh, but they weren't.

They showed frozen moments that could have been from weeks, months, even years before.

Malcolm died a few years later. Someone said his kidneys had failed.

I'm looking at Google Maps again. Malcolm's old house is white now; the garage door is closed and the driveway empty save for a flattened newspaper near the street. The grass is all dead but next door the rusty cars and trucks are gone, replaced by a red minivan parked under a pecan tree. I am haunted by something more intimate. We were once three souls together in a terrifying place, keeping each other company late at night, and all I know now is that after Malcolm died, someone painted ivy leaves on the side of his mailbox.

# Good Days and Bad

My dad and I are sitting at the kitchen table, he spooning oat bran, I sipping espresso. I'm waiting for him to pour coffee into his cereal again.

"This needs to be hot," he says, beating the plastic bowl with his spoon. "It's not hot."

Connie, the aide, asks him if he wants hot cereal. "I saw some grits in the cupboard, Doc," she says.

He says he doesn't want any damned grits.

"How're those strawberries?" I say.

"Good," he replies, looking at me over the top of his glasses. He takes another spoonful of berries and bran and chews slowly.

"Got those yesterday," I say. He sets down his spoon, picks up his cup, and goes to pour it into the bowl.

"You going to have coffee with that?" I say.

"What?" he says and stops.

"You look like you're going to pour that coffee into the bowl with the cereal and strawberries. Is that your plan?"

He looks at me again and takes a sip of coffee. Then he grins, like he's been caught in the act.

"How're those berries?" I say.

He sets down his cup.

"Good," he replies. "Where'd they come from?"

"Kroger's," I say. "We had strawberry shortcake and cream last night. Remember?"

He takes another spoonful but it's too much and milk flows down his chin.

These have become our good days. On bad days he doesn't wake up. On bad days he might lie there in his recliner with his head back and his mouth open, wheezing until something, maybe a morsel of soggy lettuce still stuck against his palate from his last supper, slides into his windpipe and he coughs and coughs until he can't go on anymore.

He's stopped eating and is looking at the black lockbox where his medications are kept.

"I need to take my pills," he says. He reaches toward the box as though summoning them.

I look at Connie and she shakes her head. "You took them this morning, Doc," she says. "Before breakfast. You said you didn't want to wait."

"Let me see that," he says and stands, trying to reach the box.

"You retired from that," I say.

"From what?"

"From being your own doctor."

"Why the Sam Hill would I do that?" he says.

"Sometimes you lose track," I say.

"The hell you say. I'm ninety-two years old, aren't I?"

I wait and he sits back down.

"Got here with no help from you."

"Yeah, well at least I know how *old* you are," I say.

"What?" he says.

"You're ninety-three. Since February," I reply, then wonder if he knows it's June.

"Shit," he says and pretends to spit. "Just tell me the god-damned combination to that box. Do something useful for a change."

In his best days, I would have trusted him to take out my gall-bladder, set my broken leg, look into my colon. Now he can't open a pack of crackers. He grips the cellophane with the sides of his fingers and his thumb, like an old man with no use for his hands anymore. I want to tell him to stop it, that he was a sur-geon, that he could sew up a man's belly with those hands faster than anyone I ever knew, that he knows how to open a pack of crackers and he should stop fucking around. Just open them, for God's sake.

I don't know why he quit surgery when he did, but I'm sure he knew it was time. Or past time. I don't think he gave up easily. Even after he retired, he still played doctor to family and friends and neighbors, often dispensing expired samples of drugs he'd hoarded for decades. A little digitalis from 1985 for a ninety-year-old woman with diabetes and emphysema, some antibiotics from 1993 for a seventy-four-year-old man with a boil on his ass, vinegar and honey for everyone poisoned by Mrs. Klein's egg salad.

In the beginning, his forgetfulness was mainly humorous, at

worst annoying. He'd call a week early to ask us if we were running late for our visit, join conversations by making up stories about events that never happened, poison a church potluck with month-old egg salad, and retain little that might cause him embarrassment or guilt.

His bad days began when he couldn't remember his medicine: what he'd taken or skipped, or why, or even which pills did what for him. We tried making a chart, taking pictures of his pills and labeling them with names and purposes, but he wouldn't use it, or he'd forget he had it and then he'd mess things up and land himself in the hospital again, sometimes with another heart attack.

I don't get angry anymore. I just watch him fumble about until he forgets what he started. "Let me get that," I say, and he gives me the crackers without a word.

This isn't the man who was my father. This man is broken and he can't fix himself and now he knows it, and so do I. Like that, bad days evolve to become the next good days.

He's almost done with his cereal and his coffee cup is empty. He's still chewing and staring out the window at the bird feeder and I wonder what he's thinking about. I guess he wouldn't know.

"Hey, Dad," I say. "Remember when you used to put barium on your cereal?"

He picks up his empty cup and tries to drink from it, looks inside, and sets it down. Connie's forehead is pinched.

"When *was* that?" I say. "Maybe a few years ago?"

"Why the Sam Hill would I do that?"

"Yes," Connie says. She has come around to the side of the table so she can see both of us. "You're making that up, right?"

"No. I'm not," I say. "He got the stuff from the radiology department, kept it in a sugar bowl right there, beside the salt and pepper. Said it prevented diverticulitis."

Dad raises his eyebrows, tilts his head. "Maybe so," he says and winks at Connie.

Connie shakes her head, picks up his spoon and empty bowl, and takes them to the sink. I'm not sure she believes Dad sprinkled barium on his cereal because he thought it prevented diverticulitis, but he did, for five or six years.

He leans forward like he's going to leave. I ask him if he thought the barium worked.

"For what?" he says and I tell him again.

He picks up his napkin and looks for his bowl and can't find it. "Don't you have any cases this morning?" he says.

"Dad, you know I don't operate anymore." I say it before I think.

He pulls himself up and squints at me. "I operated until I was sixty-nine years old," he says. "Beat that."

Connie comes back and we get him standing and turned around and walk him back to his chair in the den in front of the TV.


I was visiting at Thanksgiving a few years ago and came down with severe pain in my groin. I assumed it was my lymphoma coming back to kill me.

Dad saw me rubbing it and asked what was wrong. I said it was nothing. Nothing at all.

After dinner he found me lying in an upstairs bedroom groaning to myself. I lay on my side with my back to the door and didn't hear him come in. He asked me to roll over and I nearly jumped out of my skin. He tugged on my shoulder and I asked him to leave me alone.

"It's nothing," I said. "I'll be fine."

He told me to pull down my pants and he crowded his way onto the edge of the bed. Then he felt around, gently pushing and probing, and in less than a minute he reduced the bulge of a small hernia and made the pain go away, and in that singular moment he obliterated the utter certainty of a bone marrow transplant, interminable suffering, and a final feverish delirium.

He was ninety years old then and he couldn't remember what he'd had for breakfast, but he could still fix me with his hands.


A few days before I drove back to Ohio, I called him. Connie answered. She said he was so short of breath he probably couldn't talk.

"He's been like this for a while now," she said. "Sometimes worse."

I asked about his medicine and she said I should talk to the nurse. Or the doctor.

"It's OK," I said. "I won't tell."

She said they'd been making changes again.

"Some of the other aides claim he's agitated, but he's just Doc, you know?"

I knew.

"I'll see what I can do," I said.

I heard her take a big breath.

"Maybe I should have come out in May," I said.

"Hard to say. You can't live here."

I left a voice message for the doctor. I found the hospice nurse at her home.

"Isn't this what we expected?" she said. "*You* know, *eventually?*"

I asked about the medications. Had he been getting his diuretics? And his beta-blocker?

"Doctor made a few changes," she said.

"Like what?"

"Well, he stopped the Lasix because your dad was losing so much weight. He was getting dry, you know? And that's bad on his kidneys, which weren't in the greatest shape to begin with, you know."

"Can you give him some Lasix this afternoon?" I said. "If his weight is up, it's probably just fluid."

"He's been eating real good," she said. "Up until a day or so ago."

"So his weight is up?"

"About eight pounds," she said. "Just over the past couple of days."

She said they also stopped his beta-blocker. "His blood pressure was getting a little low, you know?"

"What's his heart rate?" I said.

"Oh, it's come up nicely," she replied. "I think it was in the eighties this morning when the nurse went by."

"Remember, he goes into failure if his heart rate goes above sixty-five," I said. "Seventy, tops."

I ask her to restart the medicines. "I'd give him a dose of Lasix now and another later in the evening," I said. "And the beta-blocker. He really needs to get his rate down, and fast."

She said she'd have to get a doctor's order. "Unless you have a license in Ohio?" she said.


I count my father's breaths. He breathes slower and slower and then he stops. I know he'll start up again, but it seems so long and Jerome, the night aide, shakes his head.

"I keep thinking this is it," he says. "But then he'll start up again and . . ."

Dad heaves a giant breath and his head jerks as he sucks great gusts of air like a steam engine. Jerome sits up straight and pulls down the front of his T-shirt.

"What makes him breathe like that?" he says.

I tell Jerome it's heart failure and he turns back to the game.

"Gets me every time," he mumbles.

Cheers erupt from the TV.

"Wish he'd seen that," Jerome says.

I've been on the road all day and my neck is in spasm and my pants are stained with cappuccino and Cheetos, but the home team just scored.

"He's been watching?" I say.

"Who? Doc? Nah, not for a while," Jerome says. "Saturday, I think. Yeah, he was awake for most of that one. Hell of a game."

"That the day he mowed the yard?"

"Yeah, it is," he says. "And worked in his shop a bit."

I'm here because I didn't believe them when they said he didn't have long, not because I did believe it.

"Did they give him some Lasix?" I say.

Jerome says he doesn't know about the morning meds.

"I'm not scheduled to give him any Lasix tonight, though," he says. "The book's in the kitchen if you want to check."

I take Dad's blood pressure, count his pulse, listen to his lungs; they are full of wet crackles and his jugular veins are full.

Jerome asks me what I think.

"I think he needs his meds," I say.

The book shows two sedatives, a sleeping pill, a pain pill, a drug for restless legs syndrome, an antipsychotic, and a diabetic pill at four times the dose he usually takes.

I had discovered that he was getting most of these drugs last week and I thought they'd agreed to stop them, especially the Zyprexa. My psychiatrist friend called it a knockout drug and said they'd have to be nuts to use it in an old man. "Unless he's threatening someone with a knife. A very sharp knife."

I don't see a diuretic or a beta-blocker on the list.

The notes mention agitation, crying out that could be pain, jerking legs and hands, and insomnia. They chart a dose of this, another of that. The drugs seem meant to erase anything in his behavior suggesting consciousness.

I open the locked medication box and take out two water pills and two doses of his beta-blocker and somehow get him to swallow them without choking.

"No other meds tonight," I say.

"What about his sleeper?" he says.

"Nothing."

I give him more Lasix, continue the beta-blocker, and hold all the crap that makes an old man crazy and shaky and sleepy and dead. I make simple changes and soon he is up and about and eating voraciously, mowing his yard again, and teasing the aides. I cook for him sometimes, help him to the bathroom, wipe up his dribbles, change his TV channels, avoid politics, and mind his medications. I get to know the five aides who rotate through his home. They've become the most important part of his life and I witness my father's gentle love for all of them as he argues and teases and complains.

I plan to leave after Father's Day. On Father's Day he becomes obsessed with his penis.

Most of the family has come to eat with him and he has had enough of the party and we're standing over his toilet, I waiting for him to remember why, and he holding on to a towel rack for balance.

"You want to go?" I say.

"Yeah. I want to go."

"Now?"

"Yeah, now. Right now."

"You want me to pull it out for you?"

"*Hell* no," he says.

He finds the zipper in his jeans and pulls it open, letting go the rack and swaying as he reaches in and pulls out his penis. He holds it for a minute and then stops. He bends over and stares at it, rolling the head between his finger and thumb.

"What the Sam Hill is this?" he says.

"Looks like a penis," I say. "Only smaller."

"There's something wrong," he says. "There's something growing on it. Right here."

I look closer. He has his foreskin wadded into a kind of ball and is rolling it around.

"It's OK," I say. "Why don't you try to pee?"

He stands for a moment, holding on to the bar again, swaying.

"What the Sam Hill is this?" he says and there he is bent over again, rolling his foreskin around. "It's some sort of tumor."

I wait, hoping he'll forget and find the will to take a piss. He pauses and stands up and I think we're getting close when he finds it again.

"What the Sam Hill is this?"

This time he asks me to take a look.

"Get me something," he says.

"What do you want?"

"What do I want? What the hell do you think I want? This thing needs to be cut out."

I take hold of his hand and gently pry it loose.

"Let me have a feel," I say. I use both hands to spread the foreskin out and ask him to take hold again.

"See?" I say. "It's gone."

"Gone?" he says. "Not my dick, I hope. I'm not done with it yet."

We left for Nebraska a few days later. I told Dad we'd be back in a month or two.

He asked me where I was going.

"Home," I said.

"Why can't I go home, too?" he said.

Back home, I call him every day or so. Only once do I catch him at a time when we make sense out of each other. I always feel worse after I hang up, and I know he forgets everything right afterward, if not before. I decide it doesn't matter whether I call or not. I decide it doesn't matter even though I know it's a common trap to think that.

It's noon on a Sunday and he'll be sitting at the head of his kitchen table, the same table where he presided over so many suppers of my childhood. I pick up the phone and it's all I can do to punch in those numbers. Our number. Our home number.

Connie answers and says he's doing pretty well.

"Ate a real good breakfast," she says. "We're not so sure about this here lunch, though."

I ask Dad about the weather and he tells me someone has stolen his car, that the cat ran away, that he had some money, a hundred dollars, but now he can't find it, that no one will let him go home and they come at night and they just sit there, waiting.

"Waiting for what?" I say.

"What do you think?" he says and grunts.

"Maybe in case you need something," I say.

"I need to go home," he says.

I'd give anything for him to be home. We've worked so hard to keep him there. I guess we didn't think he'd get so lost.

"How was lunch?"

"Hell, *I* can't get anything to eat around *here*."

"How's that turkey sandwich?" I say.

"Good," he says. "Real good."

A week goes by and the hospice nurse calls me in the evening. She says she just left him.

"I live just down Briar Avenue, you know," she says. "Sometimes I can see him when he walks by the big picture window on his way to the kitchen."

She says he's slipping. Very quickly now, she says.

I ask about his medications and learn they have stopped his Lasix again and cut his beta-blocker in half.

"He's been very agitated," she says. "So I've authorized them to use the sedatives again."

I ask about the dose and she tells me and I say it's too much, that they ought to give half that much. "Or a quarter of that," I say. "That may be what's agitating him as much as anything."

She says she understands how hard this is for me. "I'm sure you're used to making people better, even really sick people," she says. "It must be really hard to let go."

"I don't think that's fair," I say. I tell her I'm never hesitant to let go. "When it's time," I say. "When it takes too much to keep going. I understand that."

That's not where I think he is. I can't understand why this is any different from what happened a few weeks ago. I remind her about Father's Day.

"He was there at the head of the table, teasing his grand-daughters, demanding more turkey, griping about his medications again."

"Yes, I know," she says. "So much to be thankful for, isn't it?"

I tell her that the family will start gathering in Ohio on Wednesday to celebrate the Fourth.

"I'd get him back on the meds he was on when I left there right after Father's Day," I say. "And get him off all that other crap."

"These things don't always go the way that's convenient," she says. "We're here to make him comfortable, you know. Reduce his suffering."

I squeeze the phone and feel my wedding ring cutting into my finger and I want to smash something.

"He's really starting to gurgle a lot, too, Dr. Shaw," she says. "I've gotten Doctor to give me an order for some atropine to dry up his secretions."

"Those aren't secretions!" I'm yelling now. I am eight hundred miles away and suddenly my bladder feels like it's about to burst and this woman is on the phone telling me how she plans to kill my father.

"He's in failure again," I say. "That's edema fluid bubbling up in his lungs, not secretions. You give him atropine and it will probably speed up his heart rate and . . ."

And we've been here before, so I stop and switch hands and look at where my ring finger throbs.

"Just do one thing for me, OK?" I wait and she clears her throat. "Just hold off on the atropine until I get there. Please!"

She tells me again, only faster now, about how the atropine is part of their protocol, that she has been in hospice care for fifteen years and gives atropine to people all the time and that it helps a lot, and I know she has no idea what I'm talking about. I tell her I understand all that.

"But just hold off until we can get there," I say. "We'll be there by Wednesday evening. The others before that."

She never promised me anything. I think I just gave up, accepted that now might not be a bad time for it to end. I just hoped we'd get there in time.

We're a hundred miles east of Des Moines, when my brother-in-law calls to tell us Dad is dead.

"Mindy and I were with him. He was having a hell of a time catching his breath. They asked us to leave while they cleaned him up and got him dressed. We weren't out of the room ten minutes when they came out and said he'd passed."

I pull off at an exit and into a farm field.

"Official time of death was three thirty," he says.

I ask if they gave him any meds. He doesn't know. He asks how long before we arrive. I realize it's past midafternoon there and we're still half a day away and now the urgency is gone.

"We might stop in Indianapolis for the night," I say. "Neither of us got much sleep last night."

We talk about what to do with the body. I say that Dad wanted to donate his body to the medical school at Ohio State. Jeff asks if there is a document somewhere and I say I don't know. Maybe in his wallet, I say. I'm pretty sure the funeral home guy will know what to do.

He calls me back about the time we get to Peoria. Everything is taken care of, he says, but with the holiday coming up, no one at Ohio State will be there to accept the body until Monday. The funeral home Dad chose doesn't have a cooler but another one down the street does, and they have agreed to keep him over the weekend.

We are silent then and I wonder if we ought to just keep driving until we get there.

"I'm sorry I asked about meds," I say.

"The meds?" he says.

"You told me he'd died and all I could do was ask about the meds."

He says he didn't notice. "Everything was moving so fast," he says. "We didn't know what to do."


At the memorial service I see Connie and the other aides together in the pews. I sit down behind them and tell them how much Dad adored them. Connie reaches into a bag and hands me three spiral-bound notebooks.

"These are all of them," she says. "The green one is the last one."

I turn to the last page of the green one.

"Looks like he got atropine at three fifteen," I say. "Right after Mindy and Jeff stepped out."

They look at each other and then at me.

"Ten minutes," I say, and smile. "That's all it took."

I close the book and hand it to Connie. She pushes it back and tells me they're mine to keep.

I look around and see more people arriving. I wish I could stay here and talk to them awhile longer about my father and his last days.

I don't know whether Dad would have lived any longer had I stayed there to care for him.

Earlier that day, I'd bumped into the nurse who gave him the last injection. She gave me a hug and said my father had died peacefully.

"Nothing anyone could have done," she said, her hands on my arms. "His time had come."

I know she believes that, but I am still struggling to reconcile a life spent saving hopeless situations with a death that could have been so easily delayed.

# Legacy

Dad died on Wednesday, July 3. Most of the family was already coming to our annual Fourth of July reunion and we had a private gathering at the lake cottage Dad built in 1961. We spent an afternoon trading stories and laughing and crying. I thought about the story I'd sent to my sister and brother a few years before.

Yesterday I called Dad. We talked about the weather like we always do and then he asked me what I'd been doing these days. I told him again about my writing.

"Remember that story I sent you about the sigmoidoscopy in the ER?"

He didn't.

"The time you sprayed shit all over me?"

He said he'd never done anything of the kind. Then he asked if I'd done any good cases lately. I'd stopped operating almost five years before.

"I got an e-mail this morning from one of the guys I trained back in the nineties," I said. "He's president of an international surgery group and he wants me to be this year's honorary fellow."

"Oh really? Who's this?" he said.

"Not someone you've met. In his e-mail he wrote about how important I was to him and the other people he knows who trained with me. I was pretty blown away by that. What he wrote."

"That's great," he said.

"Made me feel good, anyway."

"Funniest thing happened today," he said. "I went downtown to the post office to mail something. What was it?"

"Tax returns?"

"Hell, I can't remember, but when I came out there was this old guy and his wife—well I guess it was his wife; she looked old enough but I guess it could have just as well been his sister. They were sitting in this big old Lincoln, one of those real ones."

"A Continental," I said.

"Long flat sides and a hood like a dining room table. This old boy opens the door and I didn't pay any attention but for some reason he's looking at me, waiting for me. I think he saw me when he was driving by and pulled in to wait for me to come out."

Dad stopped then.

"So was he someone you knew?" I said.

"What?" he said.

"Did you know him?"

"Oh, no. But he knew me. He came up and he had this big belly and he pulled open his shirt and showed me this scar that went from his sternum down to his . . ."

"His pubis?" I said.

"What?"

"The scar went from his sternum to his pubis," I said.

"I suppose. Couldn't really tell with his belly hanging over his belt so far. I can't remember for the life of me what I'd done, but he sure was grateful."

"So I wonder when the surgery was? It had to be at least twenty-five years, didn't it?"

"Hell if I know. I should have asked but he was so sure I'd remember him."

"Must have been more than just a gallbladder or something," I said. "With a scar that big."

"Damned if I remember," he said. I imagined Dad sitting there trying to remember some horrible surgical mess from forty or fifty years ago. "He seemed convinced I'd saved his life, though. Must have been something big. Wonder what. . . ."

"When you think about it, Dad, that's just one tiny part of your legacy. Think about all the other lives you saved or made better over those thirty-nine years. It's an incredible legacy."

"Yes," he said. "Yes, I guess it is."

I tried to think of something more to say.

"I had hopes of passing that legacy on someday," he said.

I stopped breathing.

"I guess that'll never happen," he said.

# Remission

We went back to Florida for spring break in 1963. We'd been going to Florida in the spring for as long as I could remember. Except for 1962, the year before. We didn't go then because Mom was getting cobalt treatments for her lung cancer. I remember Dad and Mom talking about taking another Florida vacation that Christmas. Mom was feeling a little better by then and she said she thought it would be a good idea. Maybe the sunshine would melt away the tumor, she said.

In years past we always drove to Florida, but in 1963 we left the station wagon at home and took a Delta DC-8 to Saint Petersburg. Mom was doing so well by then that I thought all the cobalt must have finally done some good. Nobody had said she was cured, but both Dad and Mom were sure acting differently. They seemed happy for a change, so I thought she must be all better. When the ambulance took her from the house four months later, even as sick as she was by then, I was still too much of a self-absorbed teenager to understand that would be the last time I'd ever see her.

We stayed in a motel about three blocks from the beach on Treasure Island. My brother and sister liked to swim in the warm, calm water of the pool but Mom and I liked to go to the beach and sit in the sand and watch the waves.

"Why don't you go for a swim?" she said.

I was trying to dig down to where the sand gets wet so I could build a castle or maybe a split-level with two garages. I looked at Mom and squinted against the sun.

"The waves," she said. "You can go play in the waves. You don't have to stay here with me."

"That's OK," I said. The sand was getting darker and I went back to digging.

"You need to put on some of this lotion," she said. "You're starting to look like a lobster."

Mom was wearing sunglasses with yellow frames and pink seashells on the corners. She got them in a drugstore on the other side of the bridge. Dad said she looked like Lucille Ball. Mom laughed about that. Before she got cancer she became pretty famous as a singer and dancer in the Lions Club music shows. One time she sang that song with two people arguing about how to pronounce vegetables; like "you say teh-may-teh, I say toe-mah-toe." Mom was about five feet nine and she sang it with this little guy, John, whose head came up to her elbow. On the plane ride to Florida, I heard her tell Dad about a skit she and John were going to do for the show that fall.

"Do you have to do that right here?" Mom said. "You're getting sand all over the towels."

I stood up and shook my towel off.

"Stop it! It's blowing all over me!"

"Sorry." I went back to my digging.

"Why don't you do that somewhere else?"

"But I'm almost down to the wet stuff," I said. "You know, for building a fort or something."

"Well, go down there where the tide just went out. It's *all* wet down *there*."

I had a nice wet handful of sand and I held it up and showed it to her.

"Why are you so mean to me?" she said. She stood up and took her towel and her basket of junk, moved about twenty steps away, and spread it all out again.


Dad liked taking us to the seafood restaurant at the bottom of the causeway bridge. They served crab cakes inside a horseshoe crab shell like the ones on the beach. Dad seemed happier than usual so I asked if I could have lobster. They had a plate with two Florida lobsters, french fries, and cole slaw for three dollars. Dad said they weren't real lobsters.

"More like crayfish," he said. He never let me order lobster at a restaurant, but I did it twice that trip and he didn't even argue.

One day he decided we should go waterskiing.

"In the ocean?" I said. I was worried about the big waves.

He said we'd rent a boat and skis over in the canal. "The Intracoastal Waterway," he said. "You can ski all the way to Crystal River."

We used to rent a house right on the water in Crystal River. That's why I wanted to be a marine biologist. The water was crystal clear and I could snorkel with sea cows. Some catfish lived in the bottom of the spring at the end of the dock. I held out my arms to show Dad how big they were. "That's a fish could eat a pony," he said.

The man at the dock said the boat came with a pair of skis.

Slalom skis were extra, he said. No problem, Dad said. He paid for two and the man put them in the boat and took Dad's money and stood there looking at Mom like something was wrong.

"OK, so now what?" Dad said.

"Well, sir, I reckon you want to go boating you need to put her in the water."

Dad told the man he thought he was supposed to do that.

"Boat launching? That's extra," the man said.

Mom said she wasn't feeling so good.

"Why don't you get out of the sun?" Dad said. "OK if she waits in the office there while we work this out?"

The man said it was against the rules.

"No customers allowed past that gate," he said. He stood with his hands in his pockets and pointed with his nose. I told Mom he looked like he could eat a pony but she didn't laugh.

Dad paid the launch fee and by the time we all loaded into the speedboat my younger brother was whining about being hungry and thirsty and Mom didn't look so good. Dad said we would all feel a lot better once we got out on the water.

Dad made me ski first but there were too many boats running up and down and the water was pretty rough. He took his turn and then asked Mom if she wanted to have a go. She laughed at first. Not a real laugh, though, and then she looked like she was crying and Dad told me to drive back toward the dock while he held on to her.

"But I want to ski again," I said. I'd found a place for Dad to ski out of the traffic, where the water was smooth like glass.

"Just drive, young man," Dad said.

Mom looked up then and blew her nose. "It's OK," she said. "Let him go again. The water looks really nice."

I skied and skied and ignored my dad's motions to quit until he finally cut the engine and I sank into the water.

"I want to go again," I said as Dad rolled the boat close by.

"Get in the boat."

"But, Dad—"

"Now!" he said.

On the way back to the dock I grumbled about how dumb it was to rent a boat for two hours and only use it for one, how the water had just gotten smooth, how he said I could ski all the way to Crystal River. He helped Mom get her stuff out of the boat and walked with her to the car, calling my brother and sister to get in the car with her while I held on to the boat. The man came out of the shed and backed the trailer down the ramp and into the water. I tried to get the boat in the middle of the trailer, but Dad grabbed the rope form me and did it himself. I thought he must be in a hurry but when the man pulled the tractor forward and the boat came out dripping water, Dad grabbed me by the arm, pulled me behind a big cabin cruiser on a cradle, and shook me hard.

"What the Sam Hill is wrong with you?" he said. I tried to pull my arm away but his hands were really strong. "Your poor mother is doing her best to enjoy this trip and all you care about is yourself, whining about this and that when anyone can see she's not well."

"I thought you said she was all better," I said.

"What are you talking about?"

"I heard you on the phone telling Grandma she was in tree-mission." He squinted at me as though I was crazy. "Or something like that," I said. "You said she was getting better every day."

"She's in remission."

He let go of my arm. He turned away and rubbed his face and stood with his back to me.

My sister and brother and I thought that whatever was going on with Mom's cancer, it was really good news. For months, Mom had been so happy for a change, and I heard Dad and her talking about the Florida trip like it was a kind of celebration of her getting better. Sitting by the motel pool one day, my sister asked me if Mom would start smoking again now that the cancer was gone. I felt a jolt, like panic.

Dad took me fishing on the beach. We waded out until the water was over my waist, and he showed me how to cast the big lure using both hands but I wasn't very good at it. I could cast a really long way with a regular spinning rod. I practiced in the backyard with a lead weight all the time. I laid hula hoops on the ground as targets. Mom's father was once the national champion spin caster and Dad told me that's how Grandpa Kinnear practiced. But in the ocean the waves kept knocking me down when I'd rear back to fling the bait out. One time I went under and let go of the rod and then I couldn't find it. Dad started yelling at me but then I stepped on the lure and it stuck in my foot, so he followed the line back to the rod.

He helped me back to the beach, sat me down in the sand, and tried getting the hook out of my foot, but it was stuck pretty deep.

"Guess I'll have to show you my secret trick," he said. He cut a piece of line and looped it around the curve of the hook so that he could pull the line tied to the shank one way while pulling the hook itself the other.

"See how by pulling these two lines in opposite directions, it will allow the barb to rotate down and then out when I pull this way?" he said.

"I guess," I said.

He told me to hold still. "On the count of three," he said. He held the two lines out tight, then suddenly jerked on the one around the curve and the hook just popped out.

Years later, when I worked at the clinic in Yellowstone Park, that's how I removed fishhooks from tourists' arms and legs and faces.

"That didn't even hurt," I said.

Dad said we'd wash it out good back at the house and put a bandage on it. He started packing up the tackle and unstringing the rods.

"Is Mom going to die?" I said.

Dad took the reel off his rod and put it in the box without looking at me. Then he paused with his hands resting on his knees and looked out at the waves.

"A remission means we can't find the cancer anymore, but that we don't know if it's coming back or not," he said. He dried his feet slowly, brushing sand off his soles and running a finger between his toes.

"It could come back?" I said.

"Yes."

"Even if she doesn't take up smoking again?"

"Probably."

He stood up and grabbed his tackle box. He wore this fishing cap with a bill so long he was always bumping it into things when he turned his head.

"She should have stopped before she got cancer," I said.

"Grab that rod, OK?"

We had to wait at the traffic light to cross the street.

"Can you do me a favor?" he said. "Can you be nice to your mother the rest of the time we're together? She's putting up a good show for us, but she doesn't feel as good as she acts and you're not helping matters. You understand?"

Our rooms were on the other side of the pool and I was hoping my sister and brother would be swimming, but the sun was already gone behind the building. Dad stopped me by the lifeguard chair and leaned over close.

"And for hell's sake, don't ever ask her why she didn't quit smoking, OK?"

His hand lay on my shoulder and I twisted my foot to see where the hook had stuck in me.

"That's just cruel. Do you understand that? How mean that is?"

I pressed my cut foot into the solid edge of the pool and felt the release of a sharp pain shooting through the sole.